Lecture Notes in Computer Science 14091

Founding Editors

Gerhard Goos
Juris Hartmanis

The series Lecture Notes in Computer Science (LNCS), including its subseries Lecture Notes in Artificial Intelligence (LNAI) and Lecture Notes in Bioinformatics (LNBI), has established itself as a medium for the publication of new developments in computer science and information technology research, teaching, and education.

LNCS enjoys close cooperation with the computer science R & D community, the series counts many renowned academics among its volume editors and paper authors, and collaborates with prestigious societies. Its mission is to serve this international community by providing an invaluable service, mainly focused on the publication of conference and workshop proceedings and postproceedings. LNCS commenced publication in 1973.

Pierrick Legrand · Arnaud Liefooghe ·
Edward Keedwell · Julien Lepagnot ·
Lhassane Idoumghar · Nicolas Monmarché ·
Evelyne Lutton
Editors

Artificial Evolution

15th International Conference, Évolution Artificielle, EA 2022
Exeter, UK, October 31 – November 2, 2022
Revised Selected Papers

 Springer

Editors
Pierrick Legrand
University of Bordeaux
Talence, France

Edward Keedwell
University of Exeter
Exeter, UK

Lhassane Idoumghar
Université de Haute Alsace
Mulhouse, France

Evelyne Lutton
INRA
Thirverval-Grignon, France

Arnaud Liefooghe
University of Lille
Lille, France

Julien Lepagnot
Université de Haute-Alsace
Mulhouse, France

Nicolas Monmarché
Ecole Polytechnique de Université de Tours
Tours, France

ISSN 0302-9743   ISSN 1611-3349 (electronic)
Lecture Notes in Computer Science
ISBN 978-3-031-42615-5   ISBN 978-3-031-42616-2 (eBook)
https://doi.org/10.1007/978-3-031-42616-2

This Springer imprint is published by the registered company Springer Nature Switzerland AG
The registered company address is: Gewerbestrasse 11, 6330 Cham, Switzerland

Paper in this product is recyclable.

Foreword

I was delighted to be asked to be one of the plenary speakers at EA 2022, and to have had the opportunity to share some of the exciting work that is currently going on in the field of Evolutionary Robotics with the audience.

I am not a roboticist by training. In fact, I spent many years developing algorithms for solving combinatorial optimisation problems such as scheduling and timetabling, for example in the field of hyper-heuristics. However, through interactions with various colleagues at conferences, I began to realise that robotics offered the ideal platform to investigate one of the topics in the field of Evolutionary Computation that I feel most passionate about: developing evolutionary-inspired systems that continually learn, i.e. improve their performance over time either by learning from experience and/or being able to adapt to changes in their operating environment.

I believe that the ability to continually learn is a necessary feature of many domains. In most real-world situations, the characteristics of the problems that an optimiser has to deal with are likely to change over time, as the requirements of users change, whether this is over weeks, months or years. In addition, it seems obvious that if a system is solving many instances, then it should able to learn from its experience to improve its own software. While this is true in many real-world optimisation domains (e.g. routing, factory scheduling) it seems even more acute in robotics: in addition to being asked to accomplish *tasks* that change over time, robots often operate in complex, noisy environments in which it is often challenging to foresee before deployment what obstacles or situations they might encounter. Furthermore, their operation can also be complicated by hardware failure (for example, a wheel or leg breaking) which requires a robot's control system to adapts its controller to counter the failure. As a result, robotics provides the perfect platform to be able to study methods of continual learning, particularly as learning can be applied to adapting both morphology (the robot's body) and control, thereby increasing the scope of what can be learnt over time.

Hence, I found myself somewhat unexpectedly drifting into evolutionary robotics (ER) — the subject of my talk at EA 2022. Although researchers from the field of Evolutionary Computation (EC) have applied EC to robotics for over two decades, the majority of this work has been applied to evolution of controllers. However, back in 1994, Sims [5] first proposed the use of EC to simultaneously evolve body and control of robots in simulation, with pioneering work by Lipson [3] demonstrating that complete robots evolved in simulation could be built post-evolution. Today, the opportunities to apply EC to the problem of jointly evolving body and control of robots in dynamically changing environments are plentiful. This is driven by multiple factors. Firstly, rapid advances in materials science, specifically in the development of soft materials that open up an entirely new space to evolve in, recently facilitating ground-breaking work in evolving robot designs that were subsequently built from living cells [1]. Secondly, the relatively recent sub-field of EC known as Quality Diversity (QD) is driving much of the algorithmic development in ER, through development of methods that leverage diversity to provide efficient exploration of complex search spaces [2, 4]. Finally, the ability to rapidly prototype designs via 3d-printing promises to deliver physical robots that can be easily evaluated in the real world, helping address the infamous reality gap. Given these fast-paced developments, ER seems the ideal place to study methods of continual learning in a challenging physical substrate — I'm excited about what the field might deliver in the next decade as technology, materials and algorithms advance.

Finally, writing this in July 2023 provides an opportune moment to reflect more generally on the field of artificial evolution, noting that 2023 marks the 30-year anniversary of the first ever journal in the field, Evolutionary Computation (ECJ), published by MIT Press — the first issue was published in Spring 1993. I think this clearly indicates that in 2023, Evolutionary Computation can be considered a mature science. I've had the privilege to be the Editor-in-Chief of ECJ for the last seven years, giving me oversight of the large and diverse breadth of work that is going on. I am continually surprised at the constant innovation and progress being made. The field is becoming increasingly interdisciplinary, leveraging ideas from machine learning and operations research, which is only strengthening it. I fully expect this trend to continue. The most recent issue of ECJ (volume 31:2) contains a series of articles written by the authors of the papers that were published in the first issue, reflecting on just how much the field has changed in 30 years: I look forward to seeing what a similar issue in 2053 will say!

Emma Hart

References

1. Kriegman, S., Blackiston, D., Levin, M., Bongard, J.: A scalable pipeline for designing recon_gurable organisms. Proceedings of the National Academy of Sciences **117**(4), 1853–1859 (2020). https://doi.org/10.1073/pnas.1910837117, https://www.pnas.org/content/117/4/1853
2. Lehman, J., Stanley, K.O.: Novelty search and the problem with objectives. In: Riolo, R., Vladislavleva, E., Moore, J. (eds.) Genetic Programming Theory and Practice IX. Genetic and Evolutionary Computation. Springer, New York (2011). https://doi.org/10.1007/978-1-4614-1770-5_3
3. Lipson, H., Pollack, J.B.: Automatic design and manufacture of robotic lifeforms. Nature **406**(6799), 974–978 (2000)
4. Mouret, J.B., Clune, J.: Illuminating search spaces by mapping elites. arXiv preprint https://arxiv.org/abs/1504.04909 (2015)
5. Sims, K.: Evolving 3D morphology and behavior by competition. Art. Life **1**(4), 353–372 (1994)

Preface

This LNCS volume is made of the best papers presented during the 15th Biennial International Conference on Artificial Evolution, EA[1] 2022 held in Exeter (UK). This conference is the first in the series to be held in the UK, with previous iterations held across France in Mulhouse (2019), Paris (2017), Lyon (2015), Bordeaux (2013), Angers (2011), Strasbourg (2009), Tours (2007), Lille (2005), Marseille (2003), Le Creusot (2001), Dunkerque (1999), Nimes (1997), Brest (1995) and Toulouse (1994).The conference was due to take place in 2021 but was postponed due to the ongoing effects of the COVID-19 pandemic.

We sought original contributions relevant to Artificial Evolution, including, but not limited to: evolutionary computation, evolutionary optimization, coevolution, artificial life, population dynamics, theory, algorithmic and modeling, implementations, application of evolutionary paradigms to the real world (industry, biosciences...), other biologically-inspired paradigms (swarm, artificial ants, artificial immune systems, cultural algorithms...), memetic algorithms, multi-objective optimization, constraint handling, parallel algorithms, dynamic optimization, machine learning and hybridization with other soft computing techniques. We received high-quality submissions spanning many of these areas, including theoretical advances in tree-based methods, swarm intelligence and multi-objective evolutionary algorithms in addition to a number of topical fields of application in machine learning, electric vehicles, routing and bioinformatics.

Each submitted paper was reviewed by members of the International Program Committee and selections were based on a minimum of two such single-blind reviews. This volume presents a selection of the best papers presented at the conference as in previous years (see LNCS volumes 1063, 1363, 1829, 2310, 2936, 3871, 4926, 5975, 7401, 8752, 9554, 10764, 12052).

The success of EA 2022 was the result of team work and I would like to express my gratitude to:

- Dr. Mathias Kern (BT plc) and Prof. Emma Hart (Edinburgh Napier University) for agreeing to give keynote talks;
- The Program Committee members for their rigorous work: the high quality of the selected papers demonstrates their attention to detail;
- The Organizing Committee for their efficient work and kind availability, in particular the local team;
- The members of the Steering Committee for their valuable assistance;
- Pierrick Legrand for the administration of the conference website;
- Lhassane Idoumghar for financial administration;
- Laetitia Jourdan and Patrick Siarry for publicity;

---

[1] As for previous editions of the conference, the EA acronym is based on the original French name "Évolution Artificielle".

- Pierrick Legrand, Arnaud Liefooghe, Julien Lepagnot and Nicolas Monmarché for managing submissions and for the editing of the proceedings;
- Nick Davies and Ben Samuels for their assistance in the local arrangements.

I would also like to take this opportunity to thank the following partners whose support was instrumental in delivering the conference: the Faculty of Environment, Science and Economy at the University of Exeter, Event Exeter, the University of Exeter and Association EA.

Finally, we are as always deeply grateful to all authors who submitted their research work to the conference, and to all attendees who made the conference such a vibrant venue for the exchange of ideas. The combination of scientific quality and the convivial atmosphere of this series of conferences provides a stimulating and inclusive environment for all evolutionary algorithm researchers.

Edward Keedwell
EA 2022 Chair

Évolution Articielle 2022 - EA 2022

31st October – 2nd November 2022
Exeter, UK
15th International Conference on Artificial Evolution

Chair

Edward Keedwell — University of Exeter, UK

Steering Committee

Lhassane Idoumghar	University of Haute Alsace, France
Pierrick Legrand	University of Bordeaux, France
Arnaud Liefooghe	University of Lille, France
Evelyne Lutton	INRAE Palaiseau, France
Nicolas Monmarché	University of Tours, France

Organizing Committee

Jacq Christmas (Local Organization)	University of Exeter, UK
Tinkle Chugh (Local Organization)	University of Exeter, UK
Richard Everson (Local Organization)	University of Exeter, UK
Jonathan Fieldsend (Local Organization)	University of Exeter, UK
Ayah Helal (Local Organization)	University of Exeter, UK
Lhassane Idoumghar (Treasurer)	University of Haute Alsace, France
Laetitia Jourdan (Publicity)	University of Lille, France
Pierrick Legrand (Proceedings, Website)	University of Bordeaux, France
Julien Lepagnot (Submissions)	University of Haute Alsace, France
Arnaud Liefooghe (Proceedings, Booklet)	University of Lille, France
Patrick Siarry (Publicity)	University of Paris-Est Creteil, France

Programme Committee

Hernán Aguirre	Shinshu University, Japan
Christian Blum	Artificial Intelligence Research Institute, Spain
Stephane Bonnevay	University of Lyon 1, France
Amine Boumaza	University of Lorraine, France
Mathieu Brevilliers	University of Haute Alsace, France
Stefano Cagnoni	University of Parma, Italy
Francisco Chicano	University of Málaga, Spain
Jacqueline Christmas	University of Exeter, UK
Tinkle Chugh	University of Exeter, UK
Maurice Clerc	Independent Scholar, France
Pierre Collet	University of Strasbourg, France
Laurent Deroussi	University of Clermont-Ferrand, France
Clarisse Dhaenens	University of Lille, France
Marco Dorigo	Université Libre de Bruxelles, Belgium
Rachid Ellaia	Mohamed V-Rabat University, Morocco
Andries Engelbrecht	University of Pretoria, South Africa
Francisco Fernandez de la Vega	University of Extremadura, Spain
Cyril Fonlupt	University of the Littoral Opal Coast, France
Edgar Galvan	Trinity College Dublin, Ireland
Mario Giacobini	University of Turin, Italy
Adrien Goëffon	University of Angers, France
Frédéric Guinand	University of Le Havre, France
Jin-Kao Hao	University of Angers, France
Ayah Helal	University of Exeter, UK
Lhassane Idoumghar	University of Haute Alsace, France
Anne Jeannin-Girardon	University of Strasbourg, France
Laetitia Jourdan	University of Lille, France
Edward Keedwell	University of Exeter, UK
Pierrick Legrand	University of Bordeaux, France
Julien Lepagnot	University of Haute Alsace, France
Arnaud Liefooghe	University of Lille, France
Manuel López-Ibáñez	University of Manchester, UK
Nuno Lourenço	University of Coimbra, Portugal
Evelyne Lutton	INRAE, France
Virginie Marion-Poty	University of the Littoral Opal Coast, France
Nicolas Monmarchè	University of Tours, France
Gabriela Ochoa	University of Stirling, UK
Ammar Oulamara	University of Lorraine, France
Luís Paquete	University of Coimbra, Portugal
Eduardo Rodriguez-Tello	CINVESTAV, Mexico

Frederic Saubion	University of Angers, France
Oliver Schütze	CINVESTAV, Mexico
Patrick Siarry	University of Paris-Est Creteil, France
Thomas Stützle	Université Libre de Bruxelles, Belgium
El-ghazali Talbi	University of Lille, France
Dirk Thierens	Utrecht University, The Netherlands
Alberto Tonda	INRAE, France
Leonardo Trujillo	Instituto Tecnológico de Tijuana, Mexico
Sébastien Verel	University of the Littoral Opal Coast, France
Jonathan Weber	University of Haute Alsace, France
Nicolas Zufferey	University of Geneva, Switzerland

Artificial Evolution

ea2022.inria.fr

15th International Conference
on Artificial Evolution

Invited Talks

An Evolutionary Approach to the Autonomous Design and Fabrication of Robots in Unknown Environments

Emma Hart

Edinburgh Napier University, UK

Abstract. Robot design is traditionally the domain of humans — engineers, physicists, and increasingly AI experts. However, if the robot is intended to operate in a completely unknown environment (for example clean up inside a nuclear reactor) then it is very difficult for human designers to predict what kind of robot might be required. Evolutionary computing is a well-known technology that has been applied in various aspects of robotics for many years, for example to design controllers or body-plans. When coupled with advances in materials and printing technologies that allow rapid prototyping in hardware, it offers a potential solution to the issue raised above, for example enabling colonies of robots to evolve and adapt over long periods of time while situated in the environment they have to work in. However, it also brings new challenges, from both an algorithmic and engineering perspective.

The additional constraints introduced by the need for example to manufacture robots autonomously, to explore rich morphological search spaces and develop novel forms of control require some re-thinking of "standard" approaches in evolutionary computing, particularly on the interaction between evolution and individual learning. Some of these challenges are discussed and some methods to address them that have been developed during the ARE project. Finally, some ethical issues associated with the notion of autonomous robot design, and discuss the potential of artificial evolution to be used as a tool to gain new insights into biological evolving systems is discussed.

Bio. Professor Emma Hart has worked in the field of Evolutionary Computing for over 20 years on applications ranging from combinatorial optimisation to robotics, where the latter includes robot design and swarm robotics. Her current work is mainly centred in Evolutionary Robotics, bringing together ideas on using artificial evolution as a tool for optimisation with research that focuses on how robots can be made to continually learn, improving performance as they gather information from their own or other robots' experiences. The work has attracted significant media attention including recently in the New Scientist, and the Guardian. She gave a TED talk on this subject at TEDWomen in December 2021 in Palm Springs, USA which has attracted over 1 million views since being released online in April 2022. She is the Editor-in-Chief of the journal Evolutionary Computation (MIT Press) and an elected member of the ACM SIG on Evolutionary Computing. In 2022, she was honoured to be elected as a Fellow of the Royal Society of Edinburgh for her contributions to the field of Computational Intelligence.

Optimisation Challenges at BT

Mathias Kern

British Telecommunications, UK

Abstract. In this talk, optimisation challenges from across BT Group plc, the UK's leading communications services company, are presented. Managing and operating its network and providing fixed voice, mobile, broadband and TV and products and services over converged fixed and mobile is a truly scale problem. It is discussed how optimisation approaches have been used in the design and operation of BT networks that span the whole of the UK, and the key role that they play in the daily management of 30,000 fixed resources. A number of example case studies are explored, and particular focus is given to the real-world aspects of the optimisation problems and the choice of optimisation algorithm for each particular scenario.

Bio. Dr. Mathias Kern received his MSc and PhD in Computer Science from the University of Essex, UK, in 1998 and 2006, respectively. He is currently Senior Research Manager for sustainable resource management and optimisation in the Applied Research team of BT, UK. He is an experienced industrial researcher and strong advocate for both Artificial Intelligence and Operational Research technologies and the way they interact and can be applied to real-life problems, with a particular focus on sustainable operations to help BT achieve its net-zero ambitions. He is an active member of the Operational Research Society and The Charted Institute for IT (BCS) and represents BT on the OR Society's Analytics Development Group, the Heads of OR and Analytics Forum and the BCS Specialist Group on Artificial Intelligence committee.

Contents

On the Active Use of an ND-Tree-Based Archive for Multi-Objective Optimisation

Jonathan E. Fieldsend[✉][ID]

University of Exeter, Exeter, UK
J.E.Fieldsend@exeter.ac.uk

Abstract. A number of data structures have been proposed for the storage and efficient update of unbounded sets of mutually non-dominating solutions. The recent ND-Tree has proved an effective data structure across a range of update environments, and may reasonably be considered the state-of-the-art. However, although it is efficient as a *passive* store of non-dominated solutions — which may be extracted at the end of an optimisation — its design is ill-suited to being an *active* source of parent solutions to directly exploit during a optimisation run. We introduce a number of modifications to the construction and maintenance of the ND-Tree to facilitate its use as an active archive (source of parents) during optimisation, and compare and contrast the run-time performance changes these cause (and discuss their drivers). Illustrations are provided with data sequences from a tunable generator and also a simple evolution strategy — but we emphasise such data structures are optimisation algorithm agnostic, and may be effectively integrated across the range of evolutionary (and non-evolutionary) optimisers.

Keywords: Data structures · Real-time statistics · Real-time analysis · Computational efficiency

1 Introduction

In the 1990s evolutionary multi-objective optimisation (EMO) took a large step forward with the realisation that a set of non-dominated solutions could be effectively exploited during the search process. This was enabled either through a bounded-size secondary archive or by preferentially selecting non-dominated solutions identified during search for preservation in a single search population (see e.g. [7,10,21]). Thus, algorithms started maintaining an *approximation* to the Pareto set during their search. However, it was also recognised that a *passive* archive representing the best approximation to the Pareto front found over the course of an optimisation run was often also required [20], which by its nature needed to be unbounded in size.[1] This is because relying on a bounded size approximation (or the best solutions in a final search population) typically

---

[1] In some practical examples where evaluation is very cheap this may not be feasible if the approximation set cardinality is vast, though, as shown here, modern data structures can comfortable deal with archive sizes of multiple hundreds of thousands.

© The Author(s), under exclusive license to Springer Nature Switzerland AG 2023
P. Legrand et al. (Eds.): EA 2022, LNCS 14091, pp. 1–14, 2023.
https://doi.org/10.1007/978-3-031-42616-2_1

leads to the return of a set of solutions which may be mutually non-dominating amongst themselves, but which include members who are in fact operationally worse (dominated) when all solutions visited in the search process are considered. This was shown theoretically in [9] and empirically in [6] for a size bounded archiver – with the approximation shown to be moving *backwards* and shrinking (losing extreme solutions) over time. Later work (e.g. [15]) has shown similar issues exhibited across many other bounded archiving algorithms.

Related to the above, recent work on comparing EMO algorithms has shown that contrasting optimiser performance in terms of the non-dominated set of *all* solutions visited during a search can lead to different algorithm rankings than when comparing the final populations or bounded size approximation sets [19], and as such an unbounded archive is a better reflection of an optimiser's search *capability*. However, such comparison necessitates the storing of a potentially large passive archive, which specialised data structures are much better suited to than, e.g., a simple linear list.

Many data structures have been developed to support the storage of an unbounded approximation set. These include Dominated and Non-Dominated Trees [4,6]; Quad Trees 1-3 [16,17]; Dominance Decision Trees [18]; Bi-objective Trees (or Mak_ Trees) for two-objective problems [1]; the M-front [3]; the BSP-Tree [8]; and most recently the Non Dominance Tree (ND-Tree) [11] (for which a variant has also been recently developed with re-balancing [13]). We have compared many of these data structures empirically over a range of point generation scenarios [5], and found the ND-Tree to consistently equal or better the run-time performances of the alternatives — on two very different computational architectures — often realising several orders-of-magnitude improvements in run-time. *However*, the use of unbounded archives in an active rather than passive fashion (e.g. as a source of parents in evolutionary computation to help drive the optimisation process itself) is much less explored, and typically bounded size archives are still employed for this.

In this work we are concerned with modifying the ND-Tree in order to make it amenable for use as an active rather than passive unbounded archive, and demonstrate the run-time performance changes that such necessary modifications cause. Leading from this, the main contributions of this work are:

- We identify the subroutines in the ND-Tree construction and maintenance algorithms which exhibit poor computational complexity during operations likely to occur regularly when used as a source of parent solutions as an active archive. These are the size() operation, which is recursive, and would be regularly called when sampling uniformly randomly from the set, and the maintenance of the hyperrectangle bounds (ideal and nadir estimates) for each subtree, which are loose rather than exact in the standard ND-Tree. Exact bound values are however required if sampling based on neighbourhood size, to reduce error in this weighting approach.
- We generate two versions of the ND-Tree — one which caches subtree coverage at the nodes and is therefore burdened with the additional computational cost of updating these values, but for which the call to size() becomes $\mathcal{O}(1)$; and a

Algorithm 1 Updating a non-dominated archive A with $\mathbf{x}$ (no duplicates)

Require: A	▷ The current non-dominated set of solutions
Require: $\mathbf{x}$	▷ A new solution to check against A
1: **if** $\nexists \mathbf{x}' \in A \,|\, \mathbf{x}' \preceq \mathbf{x}$ **then**	▷ If $\mathbf{x}$ is not dominated or operationally equal
2: $A := A \setminus \{\mathbf{x}' \in A \,|\, \mathbf{x} \prec \mathbf{x}'\}$	▷ Remove any members of A dominated by $\mathbf{x}$
3: $A := A \cup \{\mathbf{x}\}$	▷ Add $\mathbf{x}$ to non-dominated set
4: **return** A	

second variant where exact bounds are kept at each node, enabling the exact tracking of the minimum bounding axis-parallel hyperrectangle containing the points in the subtree rooted at a node, and its hypervolume.

- We demonstrate the empirical run-time performance of these two variants in comparison to the standard ND-Tree, both in a baseline passive archive scenario (to illustrate the additional computational cost of these changes), and in simulations of active use (where regular draws are made from the archive). These are conducted across a range of simulated data stream properties, and from simple optimiser runs.

The rest of the paper proceeds as follows. In Sect. 2 the Pareto archive updating problem is formally described, and a high-level description of the ND-Tree is presented. Note, due to space limitations we omit most low-level technical details of the data structure and its formal subroutines, but direct the reader to the original work [11]. In Sect. 3 we detail an empirical comparison of these implementations on a range of problems. The paper concludes in Sect. 4 with a discussion and highlights future work directions in this area.

2   Pareto Archive Updating

Without loss of generality, we concern ourselves with multi-objective *minimisation* problems. Given a feasible search space $\mathcal{X}$, a design $\mathbf{x}$ from this space is said to *dominate* another design $\mathbf{x}'$, which we denote as $\mathbf{x} \prec \mathbf{x}'$, if it is no worse on all m assessment criteria, $f_i(\mathbf{x})$ (i.e. $f_i(\mathbf{x}) \leq f_i(\mathbf{x}') \forall i$), and better on at least one. A design $\mathbf{x}$ is said to *weakly dominate* another design $\mathbf{x}'$, denoted $\mathbf{x} \preceq \mathbf{x}'$, if it is no worse on all assessment criteria. The set of Pareto optimal solutions (the *Pareto set*) is defined as $\mathcal{P} = \{\mathbf{x} \in \mathcal{X} \,|\, \nexists \mathbf{x}' \prec \mathbf{x}, \mathbf{x}' \in \mathcal{X}\}$. The image of $\mathcal{P}$ under $\mathbf{f}$ is known as the *Pareto Front*, $\mathcal{F}$. Note — due to many to one mappings, the cardinality of the Pareto Set may be larger than that of the Pareto Front.

As a multi-objective optimiser searches across $\mathcal{X}$ it typically maintains an approximation to $\mathcal{P}$, called its approximation set (or Pareto archive, A). The task of maintaining this set is commonly referred to as the dynamic non-dominance problem [18], and is summarised in Algorithm 1.

2.1   Archiving Limitations

Early work in the EMO field identified the various issues caused by truncating an approximation archive [6,9], and more recent research has highlighted the change in relative performance of algorithms that is observed when unbounded approximation sets are tracked [19], furthermore even quite simple sequences can induce pathological behaviours in popular (bounded size) archiving approaches [14]. As such, it is best practice to at least keep a passive unbounded archive [20], if computationally possible. From a practical point of view, even if a restricted set will be considered for presentation to the problem owner, extracting this set from an unbounded set means the problem owner is guaranteed only to consider solutions which are non-dominated by any other designs found during the optimisation run.

We now give a high-level overview of the ND-Tree and its properties.

2.2   The Non Dominance Tree (ND-Tree)

The ND-Tree of Jaszkiewcz and Lust [11] is composed of interior nodes and leaves. Each interior node has 1 to k children (usually $k = m - 1$ is chosen). Each leaf holds a set of solutions (designs). The leaf set is size bounded and breaching this size limit leads to node splitting, etc. Only leaves hold solutions, but nodes hold summary information about the subtree they root, and the ranges of the designs the subtree covers (in objective space).

The tree is constructed such that each node represents a subset of the non-dominated front which lies within a hyperrectangle defined by the subset's approximate nadir point, $\hat{\mathbf{y}}^+$, and ideal point, $\hat{\mathbf{y}}^-$, (artificial points which approximate the worst values for each objective and the best values for each objective derived from the covered point set). These are stored as node attributes. Specifically, for a node covering the set S of designs (the union of the sets of the children in the leaves of the subtree rooted at the node), $\hat{y}_i^+ \geq \max f_i(\mathbf{x}), \forall \mathbf{x} \in S$ and $\hat{y}_i^- \leq \min f_i(\mathbf{x}), \forall \mathbf{x} \in S$. Each interior node has a set of children (usually up to $m+1$), and each leave has a bucket of solutions (this capacity is user defined, but most studies use 20). The ideal and nadir are *approximate* rather than exact in the ND-Tree for computational efficiency — if a dominated solution is removed from a leaf, these bounds are not updated in the leaf (or the interior nodes which reach it), the bounds are only updated on insertion if a new point has smaller values than in the approximated ideal, or larger values than the approximated nadir in the leaf into which it is inserted (and this update is cascaded to its parent chain of nodes). There is no guarantee an inserted solution will be placed in any leaf where it instigated removal(s), and indeed a single new solution insertion may require removals in multiple leaves and/or cause removal of multiple entire subtrees to maintain the non-dominated property required in the set as a whole.

The estimated nadir and ideal locations are used to identify whether new solutions need to be compared to any of the designs covered by the node. A putative new solution $\mathbf{x}$ dominated by the nadir point of a node will be dominated by *all* members covered by that node (and so $\mathbf{x}$ can be immediately discarded). Conversely, if $\mathbf{x}$ dominates the ideal point it will dominate *all* members covered by the node (and must be accepted into the non-dominated set, and the subtree rooted at the node removed). If $\mathbf{x}$ is mutually non-dominating with respect

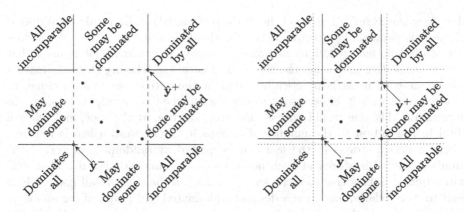

Fig. 1. Left: Illustration of a node property in the ND-Tree. A node covers a set of non-dominated points, for which the approximated ideal and nadir vectors are shown as $\hat{\mathbf{y}}^-$ and $\hat{\mathbf{y}}^+$. If a new solution $\mathbf{x}$ is found which dominates $\hat{\mathbf{y}}^-$ then the node and the subtree beneath it can be discarded. If $\mathbf{x}$ is dominated by $\hat{\mathbf{y}}^+$, then $\mathbf{x}$ can immediately be discarded. If however $\hat{\mathbf{y}}^- \not\prec \mathbf{x}$ and $\mathbf{x} \not\prec \hat{\mathbf{y}}^+$, then no further processing is needed on the node (as all points it covers must be mutually non-dominated with respect to $\mathbf{x}$). The contents of the node only need to be compared directly to $\mathbf{x}$ if none of the conditions above hold. Right: Illustration of a node property in an ND-Tree maintaining exact ideal and nadir points at each node. Note the dominates, dominated and incomparable regions are larger than that illustrated in the left example, meaning fewer putative solutions will need to be checked against the node contents. (N.B., as illustrated when $m = 2$ stored solutions lie in the corners of the tight bound axis parallel hyperrectangle which are unoccupied by $\mathbf{y}^-$ and $\mathbf{y}^+$. This is not the case in general when $m > 2$, hence regions labelled "May dominate some" and "Some may be dominated" which otherwise appear to completely dominate/be dominated when illustrated for $m = 2$.)

to both the approximated ideal *and* the nadir points, it is also mutually non-dominating with respect to all solutions covered by the corresponding node, so the subtree needs no further comparison. An illustration of these relations in objectives space for the designs covered by a node is provided in Fig. 1 (left).

As one traverses down the tree, the (hyper)volumes in objective space covered by the internal nodes decrease, and the corresponding ideal and nadir locations shift closer, until a leaf is reached containing a set of solutions residing in the objective space volume defined by the hyperrectangle defined by the leaf's ideal and nadir. These solutions all need to be compared to $\mathbf{x}$ (until the first is found which dominates $\mathbf{x}$, or the last is processed).

2.3 Modifications for an Active Use ND-Tree

The ND-Tree is designed to be efficient and effective as a store of a non-dominated set, but it is not efficient in its standard configuration as a source of parents for active use in an optimiser.

Accessing the `Size()` : Each node holds the (estimated) nadir, ideal and midpoint of the axis-parallel hyperrectangle holding the designs it covers, however it does not store *how many* designs it covers. Instead, a recursive call is used that traverses the tree and sums the set sizes in each leaf. Having this information cached at each node is practically useful in the active use of an archive, as in many scenarios it is necessary to draw a member *uniformly at random*. To accomplish this, one would traverse the tree, starting at the root, and select a child in proportion to the number of designs it covers, until a leaf is reached, where a design stored in the leaf can be picked at random. However, if the number covered is stored at each node, then any time a design is added, this will require a subroutine to increment the count in the leaf and all nodes which lead to it. Similarly, any time a design is dominated and removed (or an entire subtree removed) then this needs propagating up to all covering nodes. In our first variant of the data structure we implement such changes. Note: this also increases the memory footprint of the structure by *number of nodes × integer type memory size*. For each successful `add(x)` call there will be an extra L integer addition operations needed (where L is the number of levels above the leaf with the new design), and for each individual removal there are L integer subtraction operations needed (where L is the number of levels above the leaf with the removed design, or above the detached subtree).

Exact Nadir, Ideal and Midpoint: Having exact nadir and ideal values at nodes (and by extension an exact midpoint) also has a benefit for random sampling. When taking a uniformly random selection, no consideration is made regarding the distributional bias of the stored points. However, commonly in EMO approaches we want to have an even distribution of parents sampled across a front surface/volume, rather than biasing them in any particular objective combinations. With accurate bounds on the volumes the designs reside in (in objective space) samples can be drawn based on this volume. That is, the tree can be traversed not probabilistically based on the number of designed covered by a node, but in relation to the proportion of hypervolume lying between the ideal and nadir — i.e. probabilistically based on the volume which the designs minimally span under each node when placed in an axis-parallel hyper-rectangle. It should also be remarked that, although it brings extra maintenance cost, maintaining exact ideal and nadir points means the average number of nodes a new solution needs comparing to when evaluating whether it should be added to the set will decrease (see right illustration in Fig. 1). This means it is not immediately clear if it will be more or less costly using an exact rather than approximate nadir and ideal ND-Tree, even when using the archive exclusively in a passive scenario. In terms of modifications to the core algorithm, whenever a design is removed it needs comparing to the ideal and nadir points saved in the leaf. If it is equal on any of the values (and no other leaf member also equals them), then it defines that dimension, and it will need updating at the leaf (based on the remaining set members), and the change will need propagating up the tree until either a parent node is reached which has a smaller (ideal) or larger (nadir) on the respective index (in which case it is being derived from a different child node), or another node covered by the parent has the same minimum/maximum, or the root has been reached and updated. This can potentially require a significant number of value comparisons for a single design insertion/removal.

We now report various run-time comparisons of these implementations under different data sequence scenarios and usage scenarios. (Implementations are available at http://github.com/fieldsend.)

3   Empirical Results

We conduct our empirical work on a laptop running O/S X. The machine specifications are: 2.8 GHz Quad-Core Intel Core i7 CPU. L1 cache: 32 KB, L2 cache: 256 KB, L3 cache: 2 MB (per core), RAM 16 GB 1600 MHz DDR3.

In all experiments 30 (paired) runs are taken. Figures 2–7 show mean time taken to update a data structure to the sample/generation indicated. The shaded background on a panel indicates when a data structure with a lower mean is significantly better than another with a higher mean. Lowest blocks: lowest mean vs second lowest; middle blocks: lowest mean versus third; upper blocks: second versus third (Wilcoxon signed ranks test with Bonferronoi correction, $\alpha = 5\%$).

3.1   Simulation Runs

In our first set of experiments we employ the protocol used in [5, 8] for generation of objective vectors from controlled analytical distributions, which removes the stochastic element of the optimiser from the results. Archives are constructed from a sequence of N normally distributed objective vectors. N_d of these are dominated, and N_{nd} are non-dominated ($N = N_d + N_{nd}$). The tth objective vector $\mathbf{y}^t$ is drawn from:

$$\mathbf{y}^t \sim \mathcal{N}\left(\frac{d^t N}{t}\mathbf{1}, \mathbb{I} - \frac{1}{m}\mathbf{1}\mathbf{1}^\top\right) \tag{1}$$

where $\mathbb{I} \in \mathbb{R}^{m \times m}$ is the identity matrix and $\mathbf{1} = (1, 1, \ldots, 1)^\top \in \mathbb{R}^m$ is the vector of all ones. d^t controls the systematic improvement of points, and takes one of two values: 0 with a probability of $cN_d'^t/(N - t)$, where $N_d'^t$ is the number of dominated points still to draw in the sequence, otherwise d^t is assigned a value > 0 (here we assign 1.0). $c > 1$ results in more dominated points earlier in the sequence, and with $c < 1$ there are more dominated points later in the sequence. We investigate $c = \{0.9, 1.1\}$ here, and we measure the CPU time dedicated to the execution thread when interacting with the data structure, but exclude all other time costs (e.g., the cost of sampling from the analytical distribution).

3.2   Simulations of Usage in Passive Scenarios

In our first set of experiments we simulate usage in a passive environment: we are interested in examining the run-time cost differences in the ND-Tree implementations when they are solely being used for storage. m ranges from 3 to 20

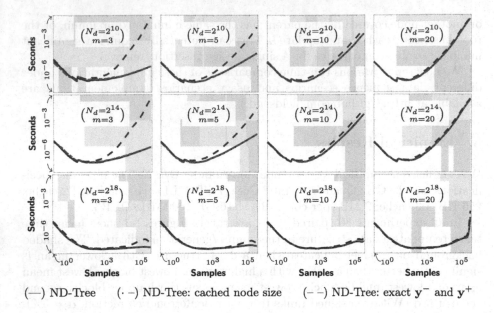

(—) ND-Tree   (· –) ND-Tree: cached node size   (– –) ND-Tree: exact $\mathbf{y}^-$ and $\mathbf{y}^+$

Fig. 2. Mean (cumulative) time taken to update data structure per sample. Results on analytical function, $c = 0.9$. Log scales on both axis.

with various configurations of the generator. All data structures get the same sequence set for each run (i.e. are *paired*), and results are shown in log–log plots.

The panels in Fig. 2 show the run-time characteristics of the different data structures for $c = 0.9$, and the panels in Fig. 3 show the run-time characteristics of the different data structures for $c = 1.1$. For both $c = 0.9$ and $c = 1.1$ and across scenario configurations the baseline ND-Tree and cached node size ND-Tree exhibit very similar run-time performances, indeed only for $m = 3, c = 0.9$ do the average performances vary substantially, but by 10^4 samples they have converged, and the different variants required between 10^{-5} and 10^{-4} seconds per sample update on average across a run.

For the ND-Tree with exact ideal and nadir values the run-time differences are substantial for fewer objectives (3 and 5) and the larger total archives sizes (i.e. where $N_d = \{2^{10}, 2^{14}\}$), outside of these though the timing differences are smaller (though still statistically significant at later stages). Where there are large variations it would appear due to the properties of these configurations – with $N_d = \{2^{10}, 2^{14}\}$ there are more non-dominated solutions in the sequences, which means more adjustments to the bounds are likely needed as solutions are more regularly added, and with the lower m sizes the trees will be deeper (as each interior node has up to $m - 1$ children, and therefore there will tend to be a longer "chain" of nodes whose $\mathbf{y}^-$ and $\mathbf{y}^+$ need updating). Nevertheless, even in the worse configurations, the average update time per sample over the run only reaches 10^{-3} seconds by the end for the ND-Tree with exact bounds at nodes.

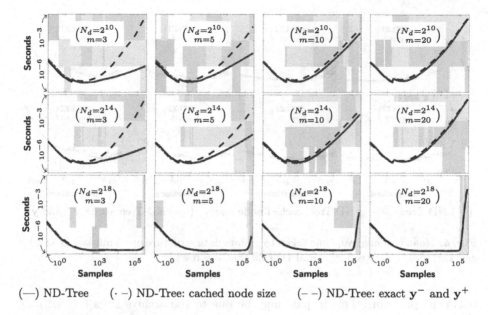

(—) ND-Tree   (· –) ND-Tree: cached node size   (– –) ND-Tree: exact $\mathbf{y}^-$ and $\mathbf{y}^+$

Fig. 3. Mean time taken to update data structure per sample. Results on analytical function, $c = 1.1$.

To illustrate the performance of the data structures in an optimisation environment, we also employ sequences generated from a simple (1+1)–Evolution Strategy (ES), as set out in [5]. The optimiser is based on the PAES algorithm of [12], but rather than using a gridded (bounded) archive, an unbounded archive is used, which is stored in the ND-Tree. The parent has a single design variable mutated with Gaussian noise, with width 0.1 (with rejection sampling for boundary violations). If the child is not weakly dominated by A, the child replaces the parent at the next generation.

We run the ES with each data structure on a problem 30 times, plotting the average update timings. DTLZ1 and DTLZ2 from [2] are used as test problems, as they usefully span two extremes in behaviour. In DTLZ1, the objective values of random design vectors are many orders of magnitude worse than those of the Pareto set. Also, the problem has many deceptive fronts — so the approximation set tends to repeatedly converge, expand, and then rapidly contract once a better local front is found. In contrast DTLZ2 is designed such that random solutions are only a couple of times worse than Pareto optimal ones on the quality criteria. Furthermore, there is a single multi-objective basin of attraction in the problem, so the approximation set tends to steadily grow over time, rather than having seismic changes in size. For both problems we set the number of design parameters as $m - 1 + 9$.

Figure 4 shows how the data structures compare – interestingly, there is little consistency in differences in the three configurations across these scenarios, with relative performance often swapping between the implementations on (and

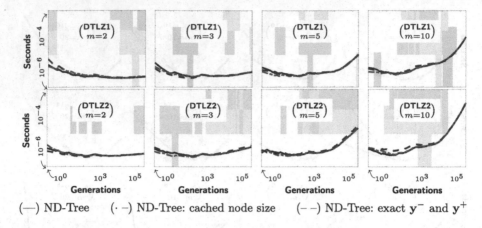

(—) ND-Tree   (· –) ND-Tree: cached node size   (– –) ND-Tree: exact $\mathbf{y}^-$ and $\mathbf{y}^+$

Fig. 4. Mean (cumulative) time taken to update data structure up to marked (1+1)–ES generation. DTLZ1 (top row) and DTLZ2 (bottom row).

between) problems. This in part may be due to the archive sizes (i.e. $|A|$). In the optimiser runs this ranged from 20,000 to 120,000 by the end (for small to large m) compared to the generator scenarios which were populating archives in excess of a quarter of a million non-dominated members by the end of a run in some configurations. The clear performance separation we see in the later stages of the analytical sequences may be greatly influenced by the ND-Tree capacity and depth reached.

3.3   Usage in Active Scenarios: Sampling at Random from the Archive

We now contrast the behaviours of the variants for sampling from the archive in a prototypical active archive scenario. In these set of experiments we interleave drawing a single member of the set with each update in the synthetic data sequence and the ES–(1+1) runs. These draws are uniformly at random from the standard ND-Tree, and the ND-Tree with cached node sizes, and in proportion to the hypervolume spanned by the nodes in the ND-Tree variant with exact ideal and nadir points tracked. Specifically, in the sequence if there are N data points checked for adding in turn to the archive, then after each "add" call to the data structure one member is drawn (without removal) from the data structure, simulating parent draws (so N draws by the end of the run). Note, we do not modify the data entry sequence for any of the runs (i.e. we do not "use" the drawn parent) — this is to ensure the computational timing results are unaffected by any induced changes in search behaviour due to particular parent draws which would change the membership of the archives being compared for each group of runs. This is because here we are purely concerned with the change in computational cost of using the data structure implementations in an "active" rather than "passive" archive setting *on the same input data sequence*.

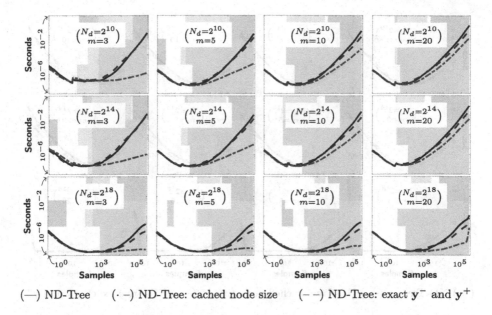

(—) ND-Tree   (· –) ND-Tree: cached node size   (– –) ND-Tree: exact $\mathbf{y}^-$ and $\mathbf{y}^+$

Fig. 5. Mean (cumulative) time taken to update data structure per sample with a draw from the archive uniformly at random at each time step. Results on analytical function, $c = 0.9$.

Results are shown in Figs. 5 and 6 for the synthetic sequence data from the generator. The time cost benefits of caching the number covered at each node compared to the standard ND-Tree when draws are regularly taken is clear across all simulation scenarios (though the relative improvement diminishes with increasing number of objectives). The variant storing the coverage number is consistently faster after 10^3 samples, ranging from 1-2 orders of magnitude speed up by the end of the run for $m = 3$ through to at least twice as fast for $m = 20$. For the version tracking the ideal and nadir points exactly, the computational cost in the active use scenario is similar to that of the baseline ND-Tree, but the samples drawn are now less influenced by a biased distribution of the archived points in objective space. Interestingly the cost of recursively calculating the node coverage in the standard configuration turns out to be similar in practice to the additionally maintenance cost of keeping the bounds exact — with the exact bound variant's run-time being very similar to the standard ND-Tree. Usually the standard approach is a little faster in the earlier stages of a run, but by later in runs the exact bound variant becomes quicker.

Figure 7 shows the results for the two DTLZ problem configurations with the interleaved draws. We see a similar trend as with the analytic functions – the ND-Tree with stored coverage attributes is significantly faster after 1,000–2,000 evaluations onwards, and reaching between 1 and 2 orders of magnitude speed up over standard ND-Tree configuration by 250,000 evaluations. The variant with

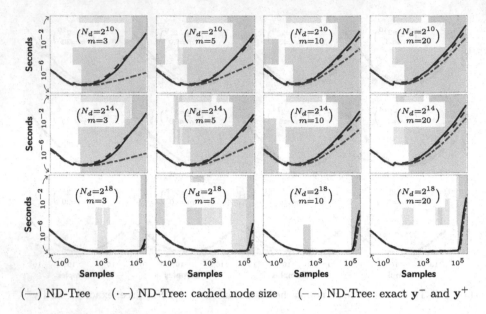

(—) ND-Tree   (· –) ND-Tree: cached node size   (– –) ND-Tree: exact $\mathbf{y}^-$ and $\mathbf{y}^+$

Fig. 6. Mean (cumulative) time taken to update data structure per sample with a draw from the archive uniformly at random at each time step. Results on analytical function, $c = 1.1$.

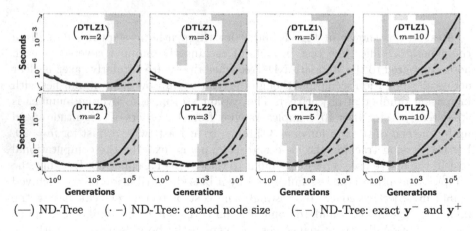

(—) ND-Tree   (· –) ND-Tree: cached node size   (– –) ND-Tree: exact $\mathbf{y}^-$ and $\mathbf{y}^+$

Fig. 7. Mean (cumulative) time taken to update data structure and draw a random member at each (1+1)–ES generation. DTLZ1 (top row) and DTLZ2 (bottom row).

exact node bounds is also more efficient that the standard data-structure, with 2–8 times speed up by the end.

4   Conclusions

In this work we set out modifications to the ND-Tree, to enable its effective use as an *active* unbounded archive in multi-objective optimisation. We compare the run-time performance of two modified variants over a range of scenarios, both synthetic, and from optimiser run histories. We find that storing the number of designs covered by each node as an extra attribute, and dynamically updating it, has only marginal run-time cost implications for the maintenance of the data structure as a whole. Furthermore, on the analytically generated sequences and optimiser histories considered it drastically improved the run-time performance of sampling uniformly from the front (often multiple orders of magnitude faster).

Storing exact nadir and ideal points at nodes rather than approximations can cause a performance degradation in some scenarios, but not punitively so – with average update costs still around 1–10ms at archive sizes in excess of quarter of a million. Storing exact values however can mitigate the distributional bias that uniformly sampling from an unbounded archive may lead to if the relative *locations* of the designs are not considered, and when used in an active rather than passive setting it ends up quicker than the standard approach. In some passive storage situations storing exact nadir and ideal points is also seen to make the data structure faster overall, but not consistently.

We look forward to exploring the impact of using unbounded archives as a source of parents in EMO runs in practical settings, and in particular investigating whether the structures will be amenable for extension to other common forms of parent draws (e.g. when using decomposition rays).

Acknowledgements. The author would like to thank the anonymous reviewers for their insightful and helpful comments.

References

1. Berry, A., Vamplew, P.: An efficient approach to unbounded bi-objective archives -: introducing the Mak-Tree algorithm. In: Proceedings of the 8th Annual Conference on Genetic and Evolutionary Computation, pp. 619–626. GECCO 2006, Association for Computing Machinery, New York, NY, USA (2006)
2. Deb, K., Thiele, L., Laumanns, M., Zitzler, E.: Scalable test problems for evolutionary multi-objective optimization. Tech. Rep. 112, Computer Engineering and Networks Laboratory (TIK), Swiss Federal Institute of Technology (ETH), Zurich, Switzerland (2001)
3. Drozdík, M., Akimoto, Y., Aguirre, H., Tanaka, K.: Computational cost reduction of nondominated sorting using the M-front. IEEE Trans. Evol. Comput. **19**(5), 659–678 (2015)
4. Everson, R.M., Fieldsend, J.E., Singh, S.: Full elite sets for multi-objective optimisation. In: Parmee, I.C. (ed.) Adaptive Computing in Design and Manufacture V. Springer, London (2002). https://doi.org/10.1007/978-0-85729-345-9_29
5. Fieldsend, J.E.: Data structures for non-dominated sets: implementations and empirical assessment of two decades of advances. In: Proceedings of the 2020 Genetic and Evolutionary Computation Conference, pp. 489–497. GECCO 2020, Association for Computing Machinery, New York, NY, USA (2020)

6. Fieldsend, J.E., Everson, R.M., Singh, S.: Using unconstrained elite archives for multiobjective optimization. IEEE Trans. Evol. Comput. **7**(3), 205–232 (2003)
7. Fonseca, C.: Genetic algorithms for multiobjective optimization: formulation, discussion and generalization. In: Proceedings of the 5th International Conference on Genetic Algorithms 1993, pp. 416–423. Morgan-Kauffman (1993)
8. Glasmachers, T.: A fast incremental BSP tree archive for non-dominated points. In: Trautmann, H., et al. (eds.) EMO 2017. LNCS, vol. 10173, pp. 252–266. Springer, Cham (2017). https://doi.org/10.1007/978-3-319-54157-0_18
9. Hanne, T.: On the convergence of multiobjective evolutionary algorithms. Eur. J. Oper. Res. **117**(3), 553–564 (1999)
10. Horn, J., Nafpliotis, N., Goldberg, D.: A niched Pareto genetic algorithm for multiobjective optimization. In: Proceedings of the First IEEE Conference on Evolutionary Computation. IEEE World Congress on Computational Intelligence, vol. 1, pp. 82–87 (1994)
11. Jaszkiewicz, A., Lust, T.: ND-tree-based update: a fast algorithm for the dynamic nondominance problem. IEEE Trans. Evol. Comput. **22**(5), 778–791 (2018)
12. Knowles, J.D., Corne, D.W.: Approximating the nondominated front using the pareto archived evolution strategy. Evol. Comput. **8**(2), 149–172 (2000)
13. Lang, B.: Space-partitioned ND-trees for the dynamic nondominance problem. IEEE Trans. Evol. Comput. **26**, 1004–1014 (2022). https://doi.org/10.1109/TEVC.2022.3145631
14. Li, M.: Is our archiving reliable? Multiobjective archiving methods on "simple" artificial input sequences. ACM Trans. Evolut. Learn. Optimiz. **1**(3), 1–19 (2021)
15. López-Ibáñez, M., Knowles, J., Laumanns, M.: On sequential online archiving of objective vectors. In: Takahashi, R.H.C., Deb, K., Wanner, E.F., Greco, S. (eds.) EMO 2011. LNCS, vol. 6576, pp. 46–60. Springer, Heidelberg (2011). https://doi.org/10.1007/978-3-642-19893-9_4
16. Mostaghim, S., Teich, J.: Quad-trees: A Data Structure for Storing Pareto-sets in Multi-objective Evolutionary Algorithms with Elitism. In: Abraham, A., Jain, L., Goldberg, R. (eds.) Evolutionary Multiobjective Optimization, pp. 81–104. Advanced Information and Knowledge Processing. Springer, London (2003). https://doi.org/10.1007/1-84628-137-7_5
17. Mostaghim, S., Teich, J., Tyagi, A.: Comparison of data structures for storing Pareto-sets in MOEAs. In: IEEE Congress on Evolutionary Computation, 2002. CEC 2002, pp. 843–848. IEEE (2002)
18. Schütze, O.: A new data structure for the nondominance problem in multi-objective optimization. In: Fonseca, C.M., Fleming, P.J., Zitzler, E., Thiele, L., Deb, K. (eds.) EMO 2003. LNCS, vol. 2632, pp. 509–518. Springer, Heidelberg (2003). https://doi.org/10.1007/3-540-36970-8_36
19. Tanabe, R., Oyama, A.: Benchmarking MOEAs for multi- and many-objective optimization using an unbounded external archive. In: Proceedings of the Genetic and Evolutionary Computation Conference, pp. 633–640. GECCO 2017, Association for Computing Machinery (2017)
20. Veldhuizen, D.A.V., Lamont, G.B.: Multiobjective evolutionary algorithms: analyzing the state-of-the-art. Evol. Comput. **8**(2), 125–147 (2000)
21. Zitzler, E., Thiele, L.: Multiobjective evolutionary algorithms: a comparative case study and the strength Pareto approach. IEEE Trans. Evol. Comput. **3**(4), 257–271 (1999)

HyTEA: Hybrid Tree Evolutionary Algorithm

Francisco Miranda⬮, Evgheni Polisciuc⬮, and Nuno Lourenço(✉)⬮

CISUC, Department of Informatics Engineering,
University of Coimbra, Coimbra, Portugal
fmiranda@student.dei.uc.pt, {evgheni,naml}@dei.uc.pt

Abstract. Hearing Loss affects an ever-growing number of people of all ages. It can occur due to a multitude of sources such as genetics, diseases, ageing, or noise exposure. If not treated properly and timely it may lead to socioeconomic difficulties such as poor job performance, hardship in finding a job, and social isolation.

In this work, we propose HyTEA, a framework based on Evolutionary Computation to create Decision Tree like models to identify people that are likely to be diagnosed with hearing loss, so they can be called for screening by a health professional. To achieve this, we will use historic data about patients who have been diagnosed with hearing problems and complement it with publicly available socioeconomic information. The models created should provide some understanding of the reason a decision is being made since this is key for health professionals.

To build Decision Trees we usually rely on greedy induction algorithms which may result in overfitting of the training data. To counter this problem, HyTEA uses a combination of two Evolutionary Algorithms, namely Structured Grammatical Evolution and Differential Evolution to generate Decision Trees.

The results show that HyTEA is capable of consistently modelling the problem space and predicting hearing loss with an accuracy of approximately 73%. Additionally, we propose a visualisation tool based on t-SNE to help identify the patients that are being wrongly classified.

Keywords: Hearing Loss · Machine Learning · Evolutionary Computation · Structured Grammatical Evolution · Differential Evolution · Decision Tree

1 Introduction

According to the World Health Organisation[1] hearing loss affects around 466 million people. By 2050 it is expected that this number doubles to around 900 million people. Of the people aged over 65, 30% are estimated to have hearing loss greater than 40 dB. The untreated patient can suffer severe social and economic consequences, greatly reducing the quality of life.

---

[1] Source: https://www.who.int/news-room/fact-sheets/detail/deafness-and-hearing-loss.

P. Legrand et al. (Eds.): EA 2022, LNCS 14091, pp. 15–28, 2023.
https://doi.org/10.1007/978-3-031-42616-2_2

There are institutions that aim to lower the severity of these consequences for example by compensating for hearing loss with hearing aids. However, for these institutions to work, it is necessary to identify and diagnose patients through regular screenings that assess the degree of hearing loss.

In this work, we propose to mitigate the impact of hearing loss in society by using Evolutionary Computation to build models that can predict if a person is likely to have a positive diagnosis, so they can be called for a hearing screening. This will allow health professionals to call potential patients for an official diagnosis, resulting in a reduction of the negative effects that the hearing impediment might bring. Given that understanding why the model is making a certain prediction is key for medical professionals, our framework relies on models that can be understandable, namely Decision Trees (DTs). Usually, to build DTs, we rely on greedy induction algorithms which might be sub-optimal, resulting in models that might become overfitted to the training data. To overcome this issue we propose the usage of a hybrid Evolutionary Computation (EC) approach based on Structured Grammatical Evolution (SGE) [9] and Differential Evolution (DE) [18]. The SGE algorithm will use grammar to specify the syntactic restrictions of the DTs, and it will be responsible for evolving their macro structure. Then, the DE algorithm will optimise the numeric parameters of each model according to the real data.

Over the years, several approaches have been proposed aiming at using EC to build Decision Trees [2, 16], most of them using Genetic Programming (GP). However, the results show that, during the evolutionary process, the population tends to be plagued with invalid individuals, which slows down the evolutionary process, compromising the overall results. To tackle this, and eliminate the occurrence of invalid individuals, we rely on Context-Free Grammar to limit the search space to a valid solution by specifying the syntax restrictions that should be followed to create DTs.

The results of our proposed approach show that HyTEA is robust, being able to consistently generate models for hearing loss prediction with accuracies above 70%, which is similar to those obtained with traditional models. The visualisation of our data with t-SNE also shows that the generated classifiers are correctly modelling the problem space.

The remainder of the paper is organised as follows. atIn Sect. 2 we showcase the key concepts required to understand the work at hand and do a brief survey of related works. Section 3 details the architecture and inner workings of HyTEA. In Sect. 4 we detail the experimental study to validate the proposed approach and in Sect. 5 we present and discuss the obtained results. Finally, Sect. 6 gathers the main conclusions.

2 Background

2.1 Evolving Decision Trees

The usage of Evolutionary Computation to evolve and design Decision Trees (DT) has been the subject of intense research. In [2], Barros *et al.* show that the vast majority of works rely on Genetic Programming (GP) using a random initialisation of trees with test values being constrained to guarantee the logical validity of the tests. Most works use the full or the *ramped-half-and-half* methods, while only one of the reviewed works uses the *grow* method. Some works use a parametric parsimony pressure approach to counter overfitting, arguing that a good balance between parsimony and accuracy is critical for efficient evolution. Works not using parsimony pressure usually do not defend this choice either since the evolution is slowed down due to larger trees and the bloat leads to overfitted DTs which do not perform well on test data.

In [16] the authors make the argument that many invalid trees are created after applying the crossover and mutation operators and that one attribute may be examined more than once along the same path from root to leaf. To tackle this issue, [16] prunes subtrees where a nominal attribute test is repeated.

2.2 Machine Learning in Audiology and Hearing Loss

There are several works that report the successful application of Machine Learning (ML) in the field of Audiology, most of them focusing on predicting a specific type of hearing loss such as noise-induced [5], sensorineural (deficiency of neural signal transfer from the cochlea to the auditory cortex) [4] and idiopathic sudden sensorineural [13].

A recent survey [3] does a review on the contributions and limitations of eight works [1,5,6,8,19–22] using ML to predict Noise Induced Hearing Loss (NIHL). The work concludes that exposure to noise above 85 dBA for over 8 h and exposure to noises over 3 kHz as the most important risk factors for NIHL. They also show that there are other factors that affect an individual's susceptibility to NIHL such as demography, hearing protection usage and mutations to genes that alter the K^+ concentration in endolymph. Most of the works surveyed used features such as age, gender, duration of noise exposure, smoking habits, working experience in years and hearing thresholds at multiple frequencies. They also were based on highly unbalanced datasets, having only between 10% to 33% of the individuals suffering from NIHL which is usually defined by patients having a hearing threshold above 25 dB. Moreover, the size of the datasets was small, with studies having sample sizes equal to or under 210, while the remaining have sample sizes of 1113, 2110 and 10567.

3   HyTEA: Hybrid Tree Evolutionary Algorithm

The goal of the proposed approach is to design Decision Trees (DT) to predict if a person is likely to have hearing problems. While aiming at maximising the predictive power of the model, we also need to balance its complexity, keeping a simple structure for high interpretability.

For this we propose HyTEA, an Evolutionary Algorithm that relies on Structured Grammatical Evolution (SGE) [9,11] and Differential Evolution (DE) [18]. The former is responsible for evolving the macrostructure of each DT, such as

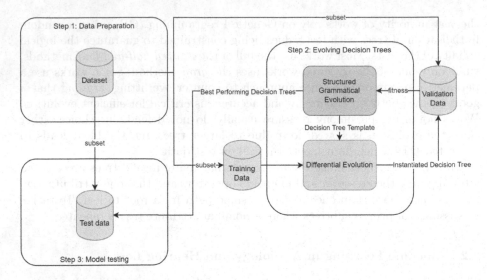

Fig. 1. Overview of the proposed hybrid architecture.

deciding the number of nodes and which features, or combination of features, should be used at each node. Each model is then passed to the DE algorithm that parses the tree and optimises the numeric parameters that will be used at each node to perform the splits. Figure 1 presents an overview of the proposed architecture.

In the first step, we prepare our dataset by performing several pre-processing operations such as feature engineering, addressing the problem of missing values and performing feature normalisation. Additionally, we split our dataset into 3 subsets: i) the Training set which will be used by the Differential Evolution component; ii) the Validation set which will be used by SGE for fitness assignment; iii) the Test set which will be used to validate the quality and generalisation ability of the best individuals found by our solution, at the end of the evolutionary search.

In the second step, we generate DTs using HyTEA. Firstly, SGE will search for the macrostructure of each model, using a grammar that defines the necessary syntax restrictions using "if-then-else" constructs as shown in Fig. 2. The symbol "%f" is a placeholder for a real number that will be searched and optimised by the DE algorithm. Using this grammar we can create DTs where a node is a leaf when the terminal symbol "is_positive(%f)" is selected to replace the non-terminal symbol <node>. Otherwise, the node will correspond to a split. In the split, a decision is done based on a condition of the form "<expr> <= %f" where "%f" is the split value of the feature calculated in <expr>. <expr>can be replaced by a numeric constant, a feature from the original dataset (represent by the x array) or a combination of features through the application of addition, subtraction, multiplication or protected division. After having the macrostructure of the DT, it is passed to the DE algorithm which will search for the numeric values of the

"%f" placeholders that maximise the prediction accuracy of the model in the Training set. Lastly, the model is evaluated in the Validation set, and it is the quality obtained in this set that will be used as fitness in SGE.

Finally, in the third and final step, the best-performing models found are used in the Test set. This step is performed when the evolutionary run is finished, to assess the generalisation ability of the best DT found. This step is paramount since it measures the extent to which the best DTs are robust and generalisable for situations beyond the training data.

```
<start> ::= <node>
<node> ::= is_positive(%f) | (<node>) if (<condition>) else (<node>)
<condition> ::= <expr><signal>%f
<signal> ::= <=
<expr> ::= <op>(<expr>,<expr>) | <var>
<op> ::= _add_ | _sub_ | _mul_ | protdiv
<var> ::= x[0] | x[1] | ... | x[60] | 1.0
```

Fig. 2. Grammar used by Structured Grammatical Evolution in the hybrid approach.

4 Experimental Setup

4.1 Dataset

To develop models capable of predicting hearing loss we built a database of 25398 patient records, with information regarding the county of origin, birth date, audiometry screenings, hearing aid usage, and responses to a Hearing Health questionnaire. The patient data was complemented with socioeconomic indicators such as demographics, education level, type of industries in the county, ageing index, salary levels, turnover per type of economic activity as well as per economic sector, the numbers of diabetes diagnosis and heart problems, the number of otorhinolaryngology exams, and fatality rates for hemorrhagic and ischemic strokes [14,15], resulting in a total of 60 features. The patient data is private and therefore can not be made accessible however all other indicators can be found in [14,15]. Indicator's data was aggregated with the calculation of means, medians, quartiles and standard deviations and features were selected via Pearson correlation.

Concerning the number of patients suffering from hearing problems, 42% of the total patients were diagnosed with loss.

4.2   Evolutionary Settings

We used the SGE implementation publicly available on Github [7]. As for the DE, we used the implementation from SciPy's Python library [17].

Since DE is sensitive to differences between the numeric values, we scaled all the features in the dataset. After some preliminary experiments with several DT models and 3 different scaling techniques, we apply a Standardisation Scaling technique bounding the features to the following interval $[-3; 3]$.

The parameters used to configure each algorithm are summarised in Table 1. The settings used by the SGE were defined following the recommendations proposed in [9,10]: {Number of Runs: 30; Population Size: 200; Generations: 100; Crossover Rate: 0.9; Mutation Rate: 0.1; Elitism: 10%; Tournament Selection with size 3; Minimum Tree Depth:3, Maximum Tree Depth: 10}. It should be noted that Maximum Tree Depth is the depth of the derivation tree, not the DT, and that once this depth is reached terminal derivations are prioritised.

For the DE algorithm, we use 15 individuals and allow the algorithm to run for 20 generations. The mutation rate is variable between 0.01 and 0.2, and we use the best/1/bin DE strategy.

Table 1. Parameters used in the experimental study for each method.

Parameter	SGE	DE
Population	200	15
Generations	100	20
Parent Selection	Tournament with size 3	N/A
Elitism	10%	N/A
Crossover Rate	0.9	0.7
Mutation Rate	0.1	Between 0.01 and 0.2
Minimum Tree Depth	3	N/A
Maximum Tree Depth	10	N/A

4.3   Fitness Assignment

Initially, the dataset is divided into three parts: 60% of the samples are used for Training, 20% are used for Validation, and the remainder 20% are used for Testing. As described in Sect. 3, the training data is used by the DE algorithm to optimise the parameters of the model. To reduce the training time, we randomly select a balanced subset of 1000 samples from the training set to be used by DE at each generation. This allows us to use all the available data for training during the evolutionary process, balancing the computational effort needed to train the model without compromising its predictive performance.

Once the individual has been optimised by the DE, we use it to classify the samples in the validation set. After all the samples are classified, we measure the accuracy of the model and use it as the fitness of the individual in the SGE algorithm.

During the parent selection stage if individuals have a validation accuracy difference lower than 2% we consider that they have the same fitness, i.e., we consider them to be tied. To resolve the ties, we take into account the individual's size measured as the number of internal nodes of the DT, i.e., individuals with fewer nodes are considered better. With this mechanism we introduce pressure towards parsimony, leading to simpler and easier-to-interpret models.

5   Results and Discussion

5.1   Training

The performance of the best individuals' accuracy over 100 generations is displayed in Fig. 3. The presented results are an average of 30 independent runs. A brief perusal of the curve reveals that HyTEA gradually improves the quality of the solutions over the entire evolutionary search. Looking at the performance of the best individuals, it is possible to see a rapid increase in the models' quality during the first 20 generations. From this point forward, the accuracy still improves, but at a much slower rate.

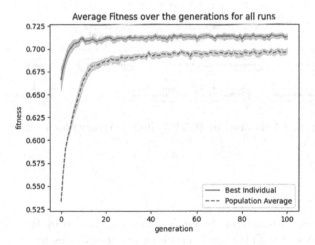

Fig. 3. Mean Best Fitness and Population Mean in the Training data across over 100 generations. The lines represent averages of 30 runs. The shadowed area represents the 95% confidence interval.

In Table 2 we present a summary of the training results for the hearing loss prediction. Looking at the quality of the best models after the 30 runs, we obtained an average accuracy of 0.72 ($\pm$0.005), with a 95% confidence interval of [0.720;0.724]. The results show a small standard deviation and a small range in

Table 2. Summary of the results for the training dataset. The results are averages of 30 runs.

	Fitness
Mean	0.722
Standard Deviation	0.005
Median	0.721
Best	0.735
95% CI	[0.720; 0.724]

the confidence interval, which indicates that HyTEA is robust, i.e., for different runs, the discovered models have a similar predictive ability.

Finally, the best-discovered model is presented in Fig. 4. This model obtained an accuracy of 0.735 in the training set. Looking at the DT, it is possible to see that it contains nodes that result from the combination of features through the application of simple arithmetic operations (e.g., the root node). This is an indication that HyTEA is also performing feature engineering when constructing the models. Table 3 details the features that are used by the best model.

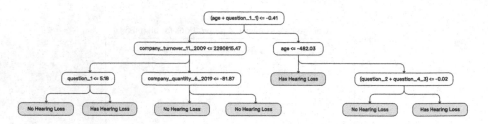

Fig. 4. Best model obtained by HyTEA after 30 runs with an accuracy of 0.735.

5.2   Testing

The absolute performance of the models that are created by HyTEA is assessed using the Test data. This step is crucial since it provides us with information about the generalisation ability of the DT. Figure 5 presents the performance of the best model discovered in each generation of the test data. The results are averages of 30 runs. Looking at the evolution of the quality of the models we can observe the same trend obtained during training, with the models gradually improving their predictive ability. At the end of the evolutionary process, the best model found has a testing accuracy of 73.7%, while the average accuracy is of 0.719 ± 0.007. The results of our experimentation are summarised in Table 4.

Another interesting outcome of this analysis is the fact that there is no evidence of overfitting since the models keep their testing performance on par with

Table 3. Description of the variables used by the best DT evolved by HyTEA (see Fig. 4).

Feature name	Description
age	Patient age
question_1_1	Answered "Yes" to question "Do you feel hearing difficulties?" (0 or 1)
company_turnover_11_2009	Total turnover of companies in the retail sector in the patient's county measured in 2009
question_1	Answer to question "Do you feel hearing difficulties?" (Between 1 and 7 where 1 is never and 7 is always)
company_quantity_6_2019	Total number of real estate companies in the patient's county of origin measured in 2019
question_2	Answer to question "Do you use a hearing aid?" (Between 1 and 7 where 1 is never and 7 is always)
question_4_3	Answer "Sometimes" to the question "Do third parties denote your hearing loss?" (0 or 1)

Table 4. Fitness, accuracy, F1, precision and recall for the 30 runs of the HyTEA experiment.

	Fitness	Accuracy	F1	Precision	Recall	Tree Depth
μ	0.722	0.719	0.721	0.716	0.727	3.262
σ	0.005	0.007	0.012	0.014	0.030	0.702
min	0.714	0.708	0.690	0.698	0.638	2.000
max	0.735	0.737	0.741	0.752	0.779	6.044

the training. This is an important result since it shows that HyTEA is not only robust but it is also able to discover models that can be used to infer and detect if a person is likely to suffer from hearing loss.

To better understand our models and the decisions that they were making, we used a visualisation tool based on the t-Distributed Stochastic Neighbour Embedding (t-SNE) [12]. This tool allows us to project all the samples in two dimensions, enabling us to study the distribution of the samples of the different classes and identify situations where our models have problems making correct predictions.

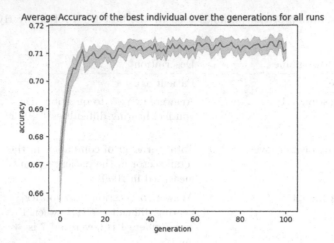

Fig. 5. Performance of the best model found in each generation on the Test Data across 100 generations. The lines represent averages of 30 runs. The shadowed area represents the 95% confidence interval.

Figure 6 presents a comparison between the original search space where each point is coloured using its true label (Fig. 6a) with the labels obtained after classifying each sample using the best DT found (Fig. 6b). In general, it is possible to see that the best DT is able to learn to distinguish between samples with hearing loss and without hearing loss. In Fig. 7 we can observe that most incorrect classifications happened in regions of high uncertainty where clusters consisted of a mix of both classes. The central cluster in the t-SNE visualisation, consisting mainly of instances with no hearing loss, had a few misclassifications. After closer inspection of the features of the samples that were wrongly classified, we found they corresponded to instances where one ear of the patient had no hearing loss whilst the other one had a severe case of hearing loss.

Lastly, we compared the performance of HyTEA against traditional ML models. In concrete, we used the Scikit-Learn framework to create tree-based models, such as simple Decision Trees, Random Forests and Gradient Boosting. In the comparisons of the models, we used the default parameters defined by Scikit-Learn, without any hyper-parameter optimisation. The average test accuracy results obtained for the simple Decision Trees, Random Forest and Gradient Boosting were 0.670 ± 0.086, 0.740 ± 0.088, and 0.741 ± 0.075, respectively. The Decision Tree and Random Forest approaches also had 100% accuracy in the training set in all runs, showing great overfitting while the Gradient Boosting approach had 76.9% accuracy in the training set, showing only slight overfitting.

When comparing to the performance of HyTEA, which obtained an accuracy of 0.719 ± 0.007, we can see that its overall performance is better than simple Decision Trees and has a similar performance to the ensemble methods while being more consistent than any of the traditional models. We applied a statistical test to compare the different approaches and found that there are meaningful

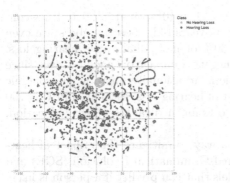

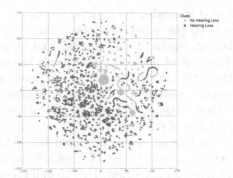

(a) t-SNE visualisation of the problem space with the correct classes.

(b) t-SNE visualisation of the problem space with the classes attributed by the best Decision Tree.

Fig. 6. t-SNE visualisation of the problem space.

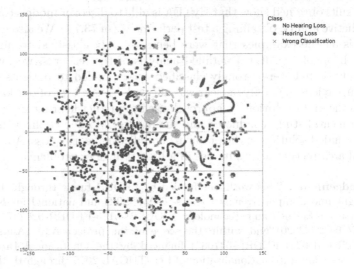

Fig. 7. t-SNE visualisation of the problem space discriminating the incorrectly classified instances by the Decision Tree. The dashed circle highlights the area of instances of partial hearing loss that were wrongly classified.

differences between HyTEA and the simple Decision Trees, i.e., HyTEA is better. Regarding the comparison with the ensemble models, there we found no differences.

6 Conclusions

Hearing loss is a health problem that is affecting an ever-growing number of people. Nowadays, around 5% of the world population suffers from hearing problems but this number is expected to grow to 10% (1 in 10 people) in 2050. If we considered the population above the age of 65 years old, these numbers are even larger, where it is estimated that over 30% of individuals have hearing loss. The lack of proper hearing levels has a negative impact on people's lives since they will have difficulties in communication. Additionally, there is an economic impact associated with the non-treatment of hearing loss, estimated in US$ 750 billion worldwide. These figures include costs for medical treatment, education, and loss of productivity.

In this paper, we proposed HyTEA, a way to mitigate the impact of hearing loss in society. HyTEA uses Structured Grammatical Evolution (SGE) and Differential Evolution (DE) to build models that can predict if a person is likely to suffer from hearing problems. Using the prediction by the models, a health official can call the person for a hearing screening to obtain an official diagnosis, which will result in a reduction of the negative effects of the hearing impediment. HyTEA uses SGE to build the macrostructure of the models, and then they are sent to a DE module so they can be further optimised.

The results obtained show that HyTEA is able to discover models that have a good predictive ability, obtaining a test performance of 73.7%. We also performed an analysis on the examples that were being wrongly classified to understand what was happening. After a detailed analysis of the features, we found that some instances that were wrongly classified corresponded to patients who had partial hearing loss, i.e., only one ear was affected. To the best of our knowledge, HyTEA is the first evolutionary approach used to predict hearing loss problems. The experimental study conducted and the results obtained confirm that it can be a well-founded solution to the problem of predicting hearing loss. Additionally, its general architecture allows its use in a broad range of problems.

Acknowledgements. This work was funded by FEDER funds through the Operational Programme Competitiveness Factors- COMPETE and national funds by FCT - Foundation for Science and Technology (POCI-01-0145-FEDER-029297, CISUC - UID/CEC/ 00326/2020) and within the scope of the project A4A: Audiology for All (CENTRO-01-0247-FEDER-047083) financed by the Operational Program for Competitiveness and Internationalisation of PORTUGAL 2020 through the European Regional Development Fund.

References

1. Aliabadi, M., Farhadian, M., Darvishi, E.: Prediction of hearing loss among the noise exposed workers in a steel factory using artificial intelligence approach. Int. Arch. Occup. Environ. Health **88**, 779–787 (2014). https://doi.org/10.1007/s00420-014-1004-z
2. Barros, R., Basgalupp, M., de Carvalho, A., Freitas, A.: A survey of evolutionary algorithms for decision-tree induction. IEEE Trans. Syst. Man Cybern. Part C Appl. Rev. **42**, 291–312 (2012). https://doi.org/10.1109/TSMCC.2011.2157494
3. Chen, F., Cao, Z., Grais, E.M., Zhao, F.: Contributions and limitations of using machine learning to predict noise-induced hearing loss. Int. Arch. Occup. Environ. Health **94**(5), 1097–1111 (2021). https://doi.org/10.1007/s00420-020-01648-w

4. Chen, X., Zhou, Q., Lan, R., Wang, S., Zhang, Y.D., Luo, X.: Sensorineural hearing loss classification via deep-HLNet and few-shot learning. Multimedia Tools Appl. **80**, 1–14 (2021). https://doi.org/10.1007/s11042-020-09702-y
5. ElahiShirvan, H., Ghotbi-Ravandi, M., Zare, S., Ahsaee, M.: Using audiometric data to weigh and prioritize factors that affect workers' hearing loss through support vector machine (svm) algorithm. Sound Vibr. **54**, 99–112 (2020). https://doi.org/10.32604/sv.2020.08839
6. Farhadian, M., Aliabadi, M., Darvishi, E.: Empirical estimation of the grades of hearing impairment among industrial workers based on new artificial neural networks and classical regression methods. Ind. J. Occup. Environ. Med. **19**, 165337 (2015). https://doi.org/10.4103/0019-5278.165337
7. Github: nunolourenco/sge3. https://github.com/nunolourenco/sge3. Accessed 29 Dec 2021
8. Greenwell, B., Tvaryanas, A., Maupin, G.: Risk factors for hearing decrement among U.S. air force aviation-related personnel. Aerosp. Med. Human Perform. **89**, 80–86 (2018). https://doi.org/10.3357/AMHP.4988.2018
9. Lourenço, N., Assunção, F., Pereira, F.B., Costa, E., Machado, P.: Structured grammatical evolution: a dynamic approach. In: Ryan, C., O'Neill, M., Collins, J.J. (eds.) Handbook of Grammatical Evolution, pp. 137–161. Springer, Cham (2018). https://doi.org/10.1007/978-3-319-78717-6_6
10. Lourenço, N., Ferrer, J., Pereira, F.B., Costa, E.: A comparative study of different grammar-based genetic programming approaches. In: McDermott, J., Castelli, M., Sekanina, L., Haasdijk, E., García-Sánchez, P. (eds.) EuroGP 2017. LNCS, vol. 10196, pp. 311–325. Springer, Cham (2017). https://doi.org/10.1007/978-3-319-55696-3_20
11. Lourenço, N., Pereira, F.B., Costa, E.: Unveiling the properties of structured grammatical evolution. Genetic Program. Evolvable Mach. **17**(3), 251–289 (2016). https://doi.org/10.1007/s10710-015-9262-4
12. Van der Maaten, L., Hinton, G.: Visualizing data using t-SNE. J. Mach. Learn. Res. **9**, 2579–2605 (2008)
13. Park, K.V., et al.: Machine learning models for predicting hearing prognosis in unilateral idiopathic sudden sensorineural hearing loss. Clin. Exp. Otorhinolaryngol. **13**, 148–156 (2020). https://doi.org/10.21053/ceo.2019.01858
14. PORDATA. https://www.pordata.pt/ (2019). Accessed 19 Dec 2021
15. Portal de Transparência do SNS. https://www.sns.gov.pt/transparencia/ (2019). Accessed 19 Dec 2021
16. Saremi, M., Yaghmaee, F.: Evolutionary decision tree induction with multi-interval discretization, pp. 1–6 (2014). https://doi.org/10.1109/IranianCIS.2014.6802543
17. SciPy: differential evolution. https://docs.scipy.org. Accessed 29 Dec 2021
18. Storn, R., Price, K.: Differential evolution – a simple and efficient heuristic for global optimization over continuous space. J. Global Optim. **114**, 341–359 (1997). https://doi.org/10.1023/A:1008202821328
19. Zare, S., Ghotbi Ravandi, M.R., ElahiShirvan, H., Ahsaee, M., Rostami, M.: Predicting and weighting the factors affecting workers' hearing loss based on audiometric data using c5 algorithm. Ann. Glob. Health **85**, 88 (2019). https://doi.org/10.5334/aogh.2522
20. Zare, S., Hasheminejad, N., Shirvan, H., Hasanvand, D., Hemmatjo, R., Ahmadi, S.: Assessing individual and environmental sound pressure level and sound mapping in Iranian safety shoes factory. Roman. J. Acoust. Vibr. **15**, 20–25 (2018)

21. Zhao, Y., et al.: Machine learning models for the hearing impairment prediction in workers exposed to complex industrial noise: a pilot study. Ear Hear. **40**, 1 (2018). https://doi.org/10.1097/AUD.0000000000000649

22. Zhao, Y., Tian, Y., Zhang, M., Li, J., Qiu, W.: Development of an automatic classifier for the prediction of hearing impairment from industrial noise exposure. J. Acoust. Soc. Am. **145**, 2388–2400 (2019). https://doi.org/10.1121/1.5096643

A Game Theoretic Decision Tree
for Binary Classification

Rodica Ioana Lung[ID] and Mihai-Alexandru Suciu[(✉)][ID]

Centre for the Study of Complexity, Babeş-Bolyai University, Cluj-Napoca, Romania
rodica.lung@econ.ubbcluj.ro,
mihai-suciu@cs.ubbcluj.ro
http://csc.centre.ubbcluj.ro/

Abstract. Decision trees are some of the most popular and intuitive classification techniques. Based on the recursive division of the data, the goal is to ultimately identify regions in the space in which most instances belong to the same class. This paper proposes a game-theoretic decision tree using a two-player game to determine the splitting hyperplane at the node level based on the Nash equilibrium concept. The entropy on each sub-node is used as a payoff function that has to be minimized. The game's equilibrium can be computed by minimizing an objective function constructed based on Nash equilibria properties. A new selection mechanism is proposed for the Covariance Matrix Adaptation - Evolution Strategy (CMA-ES) in order to approximate equilibria at each node level. Numerical experiments illustrate the behavior of the approach compared with other decision trees based methods.

Keywords: Binary classification · Nash equilibrium · CMA-ES

1 Introduction

Decision Trees (DT) and binary classification problems form an excellent combination as the former can be seen as a simple and intuitive representation of the latter's solution. They were listed early on among the top 10 algorithms for data mining [21] and have been used extensively in applications in many fields [19]. However, the simplicity of interpretation of axis-parallel ones has been replaced by various performance-enhancing splitting techniques, starting with oblique and nonlinear variants, followed by various hybridization with other methods such as neural networks, clustering, etc. Most DT induction algorithms make use at some point of the concept of optimization as they search for the best splitting mechanism in the form of the maximum/minimum of some indicator/error measure. However, in situations in which a trade-off is required, such as the one needed to avoid overfitting, optimal solutions may not be the most efficient.

When dealing with trade-off situations, game theory provides a series of solution concepts designed to overcome some of the disadvantages of the optimal solutions in conflicting situations. One of the most popular is the Nash equilibrium, which ensures stability against unilateral deviations. This paper proposes a simple method to explore the use of the Nash equilibrium concept to construct

P. Legrand et al. (Eds.): EA 2022, LNCS 14091, pp. 29–41, 2023.
https://doi.org/10.1007/978-3-031-42616-2_3

splitting hyper-planes at non-terminal node levels of a DT. A game in which the two sub-nodes try to minimize their entropy is designed. The game's equilibrium is approximated by minimizing a simulation-based objective function.

2   Decision Trees

The binary classification problem can be described as finding a rule for assigning labels to data instances based on the information provided by a training set consisting of data from the same distribution for which the labels are known. Let $\mathbb{D}$ be the given data set, $\mathbb{D} \subset \mathbb{R}^{N \times p}$, consisting of data instances $x_i \in \mathbb{R}^p$, and $y_i \in Y$, $Y \subset \{0,1\}^N$, the corresponding labels [9,22].

Decision trees (DT) [2] generate such rules by splitting data into regions that are "pure" with respect to one label, i.e., most data instances in that region have the same label; they assume that other, unlabeled instances, belonging to the same region would also have the same label. Regions are defined by using hyper-planes, either axis parallel [2] or oblique [14,20]. Nonlinear separation methods also have been proposed, for example, by [12]. Compared to axis parallel splitting methods, oblique and nonlinear ones are known to be more efficient at the expense of computational complexity. The tree representation allows the recursive division of the data space while also being intuitive in representing the defined rules.

Thus, each tree node contains some data that must be split (or not). If the data can be considered "pure enough" based on some indicator - usually derived from the proportion of instances having the same label, then that node becomes a leaf node for the tree. If not, a splitting rule separates the data into two subsets as "pure" as possible that will be assigned to its two sub-nodes. Decision trees differ in how they split node data, evaluate the purity, and perform the prediction based on leaves data.

Oblique decision trees typically test at each node an expression of the form

$$\sum_{j=1}^{p} a_j x_{ij} + a_{p+1} \leq 0,$$

where $a_1, a_2, \ldots, a_{p+1} \in \mathbb{R}$ are parameters defining the hyperplane and $x_i = (x_{i1}, x_{i2}, \ldots, x_{ip}) \in \mathbb{D}$ is a data instance. The induction of an optimal oblique DT is a computationally challenging task. Moreover, unlike in the case of axis-parallel DTs, the exhaustive search for the best splits is mostly not feasible.

There are many flavors of oblique decision trees. In [20] the authors propose HHCART as an oblique decision tree that uses Householder matrices to estimate data orientation during the split at non-terminal node levels. Another approach hybridizes a neural network with a decision tree to get the best of the two worlds: the precision of the neural network combined with the readability of the decision tree [18]. Another hybridizing technique uses fuzzy set theory and fuzzy information theory to create fuzzy ODTs [4]. In a bottom-up approach to the induction of decision trees, clustering and binary classification techniques are used to create leaves; hyper-planes are optimized by using feature selection [1]. The explainability of the oblique decision trees has also been a concern as the adoption of a classification model by non-practitioners is dependent mainly on its ease of interpretability [10,11]. The approach presented in this paper is novel because it attempts to use the Nash equilibrium concept at the node level for splitting data to take advantage of the stability properties of this solution concept. In what follows, the proposed game and decision tree model are presented.

3 ESDT - Equilibrium Split Decision Tree

ESDT uses the concept of Nash equilibrium to split node data X, y. The Nash equilibrium is one of the most popular solution concepts for non-cooperative games. A game is defined by a set of players that have to choose among a set of actions/strategies and receive a payoff based on their choices. The game aims to maximize (or minimize) each player's payoff. However, as in most situations, all the players cannot maximize their payoffs simultaneously; the compromise proposed by the Nash equilibrium concept is a situation in which no player can improve its payoff by unilateral deviation.

Consider the normal form game $\Gamma_{X,y}$ consisting of:

- **Two players** corresponding to the two sub-nodes of a parent node, denoted by $\mathcal{L}$-left and $\mathcal{R}$-right. The subset of X and y corresponding to the left sub-node is denoted by $X_{\mathcal{L}}$ and $y_{\mathcal{L}}$ respectively, and to the one to the right sub-node by $X_{\mathcal{R}}$ and $y_{\mathcal{R}}$.
- Each player chooses as a **strategy** a splitting parameter $\beta_{\mathcal{L}}$ and $\beta_{\mathcal{R}}$ for X with the aim to minimize its own entropy.
- the **payoffs** $E_{\mathcal{L}}(\beta_{\mathcal{L}}, \beta_{\mathcal{R}})$ and $E_{\mathcal{R}}(\beta_{\mathcal{L}}, \beta_{\mathcal{R}})$ are computed as the entropy of its data if X is split by using parameter β computed as the average of $\beta_{\mathcal{L}}$ and $\beta_{\mathcal{R}}$.
  In order to evaluate the entropy of the sub-nodes, the data in the node is split using parameter β in the following manner: an element $x \in X$ is placed in $X_{\mathcal{L}}$ if $x^T \beta <= 0$, and otherwise is placed in $X_{\mathcal{R}}$ (Fig. 1).

The equilibrium of this game represents a set of split parameters $(\beta_{\mathcal{L}}, \beta_{\mathcal{R}})$ such that there is no possible decrease of the entropy of either node by unilateral deviation of any of the players.

Equilibrium Split Decision Trees (ESDT) use the equilibrium of game Γ to split data at a node level. At each node level, the equilibrium of the game $\Gamma_{X,y}$ is approximated, and the data is split based on the information provided by the game. The induction of the tree stops either when a maximum depth is reached or when data in the nodes is "pure", i.e., all instances belonging to that node have the same label.

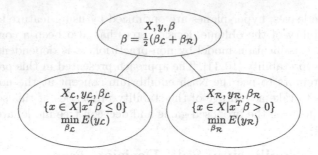

Fig. 1. Node splitting procedure in game $\Gamma(X, y)$. Nodes to the left and to the right choose parameters $\beta_{\mathcal{L}}$ and $\beta_{\mathcal{R}}$ to minimize their entropy. The parent node is split using β, the average of the two.

Node Level. At the node level, the data X, y are split based on the equilibrium strategy of game $\Gamma_{X,y}$ by using the strategy of the left player, $\beta_{\mathcal{L}}$. In this manner, the left node will have a better separation potential as $\beta_{\mathcal{L}}$ is fitted for data likely to be placed on this node. However, on the other hand, the right node will contain more mixed data values separated in the next step. In this manner, an asymmetric tree will be created in which the probability that the right nodes will be further split will be higher than that of the left nodes.

Algorithm 1. Node split

1: **Input:** X, y;
2: **Output:** $X_{\mathcal{L}}, y_{\mathcal{L}}, X_{\mathcal{R}}, y_{\mathcal{R}}$, and β to define the split rule for the node;
3: Approximate (using βCMA-ES)

$$(\beta_{\mathcal{L}}^*, \beta_{\mathcal{R}}^*) = \text{argmin}_{\beta_{\mathcal{L}}, \beta_{\mathcal{R}}} \nu(\beta_{\mathcal{L}}, \beta_{\mathcal{R}} | X, y)$$

4: Set $X_{\mathcal{L}} = \{x \in X | x^T \beta_{\mathcal{L}}^* \leq 0\}$ and
    $y_{\mathcal{L}} = \{y \in Y | x \in X_{\mathcal{L}}\}$
5: Set $X_{\mathcal{R}} = \{x \in X | x^T \beta_{\mathcal{L}}^* > 0\}$ and
    $y_{\mathcal{R}} = \{y \in Y | x \in X_{\mathcal{R}}\}$
6: **Return:** $X_{\mathcal{L}}, y_{\mathcal{L}}, X_{\mathcal{R}}, y_{\mathcal{R}}, \beta_{\mathcal{L}}^*$

Leaf Level. As there is no reason to assume that the equilibrium of the game provides a perfect separation but instead may provide good separable data for the sub-nodes, the parameter β of the probit classification estimated by using MLE (see Example 1) is computed and preserved in the leaves and used to make predictions if the leaf is not pure and contains at least two instances with a different label than the majority. If the probit model cannot be used, i.e., the leaf is 'pure' or contains only one instance of data with a different label than the rest, probabilities for test data are computed based on the proportion of instances of each class in the leaf.

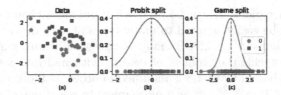

Fig. 2. A synthetic data set with 50 instances: (a) the data set X, y; (b) distribution of $X^T\beta$, where β is the probit parameter computed using MLE; (c) distribution of $X^T\beta_{\mathcal{L}}$, where $\beta_{\mathcal{L}}$ is the CMA-ES approximation of the equilibrium strategy for the left-node.

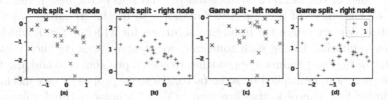

Fig. 3. Split data in each sub-node corresponding to Fig. 2(b) in (a) and (b) and to Fig. 2(c) respectively in (c) and (d).

Equilibria Computation - βCMA-ES The equilibrium of game $\Gamma_{X,y}$ is a strategy profile such that no player has an incentive for unilateral deviation. We can actually use this property to approximate a Nash equilibrium of the game. A number of unilateral deviations uniformly distributed are generated for each player and the sum of squares of the deviations that lead to a better entropy value is minimized. We denote this function by $\nu(\beta_{\mathcal{L}}, \beta_{\mathcal{R}} | X, y)$:

$$
\begin{aligned}
\nu(\beta_{\mathcal{L}}, \beta_{\mathcal{R}} | X, y) = \sum_{k=1}^{n_d} \big(&\max\{0, E(\beta_{\mathcal{L}}, \beta_{\mathcal{R}}) - E(\beta_{\mathcal{L}} + U([a, b], p), \beta_{\mathcal{R}})\}^2 \\
&+ \max\{0, E(\beta_{\mathcal{L}}, \beta_{\mathcal{R}}) - E(\beta_{\mathcal{L}}, \beta_{\mathcal{R}} + U([a, b], p))\}^2 \big)
\end{aligned}
\tag{1}
$$

where $U([a, b], p))$ denotes p uniformly distributed values in the interval $[a, b]$ and n_d is the number of deviations used for the evaluation.

Function ν simulates the corresponding optimization approach to Nash equilibria computation for normal form games [13] where a similar (exact) function has global minimas with value 0 for all the Nash equilibria of the game. Thus, minimizing ν should lead to solutions that are close to or present some equilibria properties such as stability against unilateral deviations for some of the players.

The minimization of function ν is not a trivial task, especially since we do not have an analytical form. Moreover, an additional challenge is that for some data sets, we can expect multiple solutions to be equilibria as multiple parameters classify data in the same manner. However, we can test the assumption that such an approach may be useful in designing a DT, by using an optimization heuristic such as CMA-ES: Covariance Matrix Adaptation Evolution Strategy [8] to approximate game equilibria by minimizing function $\nu()$. CMA-ES evolves the mean and covariance matrix of a population of potential solutions to the problem to find optimum regions. It can be used with minimum changes in parameter settings and can be easily adapted to many types of optimization problems.

Thus, to minimize $\nu()$, CMA-ES may be run with default parameters. However, in order to tackle the problem of dealing with multiple possible solutions the selection method of CMA-ES is modified in order to additionally ensure the minimization of $\|\beta\| = \sum_{j=1}^{2p} \beta_j^2$ as a mechanism to select one of the parameters from the set the optimal ones.

CMA-ES evolves the mean and covariance matrix of a normally distributed population. Each iteration it generates a population of individuals by using the current values of the mean covariance matrix and selects the best μ individuals - based on the values of the objective function - to use for updating population parameters. We are interested in finding solutions for which $\nu() = 0$, which may be game equilibria. Among such solutions, we want to preserve the "smallest" ones. One of the most common approaches to such a problem is to add $\|\beta\|$ to the objective function, multiplied by some factor. Here we propose a new mechanism to minimize first the objective function $\nu()$ and among optimal solutions to further minimize $\|\|$: the objective function to be minimized is $\nu()$; when more than μ individuals converge to the (known) optimum of 0, their fitness is modified in the following manner:

(i) $\|\beta\|$ is added to all individuals having the fitness value equal to 0; in this manner they will be sorted and selected based on their magnitude;

(ii) in order to avoid individuals γ with $\nu(\gamma) > 0$ to appear to be better than individuals in the first situation, the maximum of $\|\beta\|$ from all β having $\nu(\beta) = 0$ is added to their fitness.

Thus $\| \cdot \|$ is minimized only if necessary, and only solutions that are aleady optimal are subject to this minimization process. We labeled this adaptive version of CMA-ES by βCMA-ES.

Example 1. Consider the data set X, y represented in Fig. 2(a)[1]. A simple way to find a parameter β that can optimally split data (X, y) is to use *probit classification* [7]. Within this model the probability that an instance x has label 1 is estimated by using the cumulative distribution function of the standard normal distribution $\Phi(\cdot)$:

$$P(y = 1|x) = \Phi(x^T \beta), \qquad (2)$$

where β is the model parameter computed by maximum likelihood estimation (MLE). MLE optimizes the log likelihood function in an attempt to find β such that it maximizes probabilities in (2) for all $x \in X$ with label 1 and minimizes them for all $x \in X$ having label 0.

Parameter β can be used to separate data in X: if $\Phi(x^T \beta) \geq 0.5$ then the model assigns to x the label $\hat{y} = 1$ and x would be placed in the right sub-node. But the condition $\Phi(x^T \beta) \geq 0.5$ is equivalent to $x^T \beta \geq 0$ so β can be used to

---

[1] Generated by using the function: make_classification(n_samples=50, n_features=2, n_redundant=0, n_informative=2, n_classes=2, random_state=50, class_sep=0.5, weights=[0.5]) from the Python module sklearn.datasets.

define a separating hyperplane for a node. In the same manner, if $x^T\beta < 0$ then x will be assigned label $\hat{y} = 0$ and placed in the left sub-node. In fact any β value such that (ideally) for all x with label 1 we have $x^T\beta \geq 0$ and for those with label 0 we have $x^T\beta < 0$ would provide a reasonable hyperplane to split data. Figure 2(b) represents the values of $X^T\beta$ with corresponding labels for the Probit classification model distributed under the standard normal probability distribution (green). To the right, the values of $X^T\beta_{\mathcal{L}}$ are represented in the same manner, but here $\beta_{\mathcal{L}}$ is the approximation of the Nash equilibrium strategy of the first player (left node) of the game $\Gamma_{X,y}$ computed by minimizing function ν in (1) with βCMA-ES. Figure 3 represents the distribution of data in sub-nodes in each case. Corresponding entropy values for the sub-nodes are for probit: 0.82, 0.87; and for game Γ: 0.70, 0.84.

4   Numerical Experiments

Numerical experiments are performed to illustrate the potential of the proposed approach on a set of synthetic and real-world data-sets with various degrees of difficulty.

Synthetic Data-sets. In order to ensure reproducibility and also to control the characteristics of the benchmarks, we use the `make_classification` function from the `scikit-learn` library [15] to generate the data, and we report the parameters for each variant of the data. The parameters used to generate data sets are all combinations of: number of instances $(100, 250, 500)$, number of features $(2, 5, 10)$, number of classes (2), class separator $(0.1, 0.2, 0.5, 1)$, weights for each class (0.5), and random state (500).

Real-World Data. We use four real world data sets[2] that represent a binary classification problem and have various degree of difficulty. The data sets used are: iris data set (D1) from which we removed the *setosa* instances in order to obtain a linear non separable binary classification problem, Pima Indians Diabetes data set (D2) with eight attributes and 768 instances, the sonar data set (D3) with 60 attributes and 208 instances, and Haberman's Survival data set with three attributes and 306 instances (D4).

ESDT Parameter Settings. ESDT was run using the following parameters: maximum tree depth 5 and 10; CMA-ES population size 15 and maximum number of fitness function evaluations 1500; in evaluating $\nu()$ we generated 500 unilateral deviations for each player, following a uniform distribution in the interval $[-0.5; 0.5]$.

---

[2] UCI Machine Learning Repository https://archive.ics.uci.edu/ml/index.php, accessed October 2021.

Comparisons With Other Methods. We compare the performance of ESDT with the following classifiers:

- *Decision Tree* (DT) classifier [2] in two variants the use the *gini* and *entropy* indicators as a split criterion; with a maximum depth of 0 (split until all leaves of the decision tree are pure), 5 and 10 for each variant;
- *Random Forest* (RF) classifier [3] with 5 and 10 estimators respectively, the *gini* or *entropy* indicator as split criterion, a maximum depth for the estimators of 0 (split until all leaves are pure), 5 and 10;
- *Oblique Decision Tree* (Oblique DT) classifier [20] that uses MSE for the impurity criterion, split based on average of each feature, maximum depth of 5 and 10.

For the compared DT and RF classifiers we use their implementations from the `scikit- learn` library and for the Oblique DT we use the implementation available [5].

Performance Evaluation. 10-fold cross-validation is used: each data set is split into $k = 10$ folds using the `StratifiedKFold` method from `scikit-learn` (with seeds $60, 1, 2, 3, 4$) [9]. In order to estimate the prediction error each fold is used once as test data and the rest as training data. Considering all parameters, in the synthetic setting, we obtain 360 different data sets with different degrees of difficulty which is a good test-bed for classification problems [17]. For each classification model we report the AUC - area under the ROC curve [6,16] for each k test fold of each classification problem, resulting in 50 AUC values for each data set, that can be used in paired t-test comparisons between methods. AUC takes values between 0 and 1, a higher value indicate a better probability to correctly classify instances considered "true".

Results. We compare the performance of ESDT with the other methods based on AUC values reported by each classification model on the $k = 10$ test folds of each data-set. Figure 4 presents the p-values from a paired *t-test* that compares AUC values on all tested folds of ESDT against the compared models, testing the null hypothesis that ESDT results are worst than the other models. For data sets with fewer features, 2 and 5 more than for data sets with a larger number of instances, ESDT performs significantly better than the other classification models. In fact, in 49% of the cases, the p value indicates that ESDT results can be considered better and 14% of the tested data sets worse. Table 1 presents the percent of ESDT results that can be considered better/worst for the synthetic data sets compared with each of the other methods.

On the one hand the performance of ESDT does not seem to depend on the number of instances but on the number of features. However, we used the same number of fitness function evaluations for βCMA-ES for all number of attributes, so differences in performance can be expected. We find that for 10 attributes ESDT is outperformed most by random forests models, which is also to be expected. Even-more, for smaller number of features ESDT does report better results even than random forests. Also, ESDT does not use any mechanism

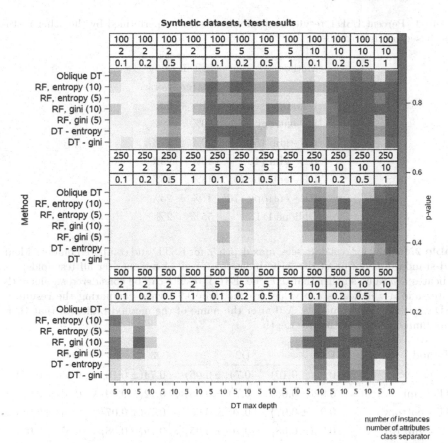

Fig. 4. Synthetic data-sets, comparisons with other methods. Colors represent p-values resulted from a *t-test* comparing AUC values reported on the 10 test folds, with the null hypothesis that ESDT results are worst than the other method. A p - value smaller than 0.05 can be considered significant, the null hypothesis rejected, and results reported by ESDT regarded as better than those reported by the other method. The first line in the headers indicates the number of instances, the second one the number of attributes, and the third the class separator.

to manage the features used to split data at node level: the advantage of this is that there is no extra selection process added to the tree induction.

Tables 2 and 3 present results reported by all methods for the real datasets when the maximum tree depth is set to 5 and 10, respectively. Mean and standard deviation of AUC values, as well as *t-test* results, are presented. The other methods are also run with maximum depth tree 0, i.e., without any limit in tree size. We find results to be similar to those observed in the case of synthetic data sets, confirming that this approach has the potential to compete and outperform other methods.

Table 1. Percent ESDT results outperformed/were outperformed by the other methods, based on *ttest* results.

Method	ESDT	
	better	worse
DT-gini	51%	6%
DT-entropy	59%	12%
RF - gini (5)	50%	13%
RF - gini (10)	40%	23%
RF - entropy (5)	48%	15%
RF - entropy (10)	37%	23%
Oblique DT	55%	2%

Table 2. Results for real data sets, max depth 5, for ESTD and compared models. Mean and standard deviation of the AUC indicator for all classifiers over all test folds; a ⋄ indicates a p value smaller that 0.05, i.e. ESDT results can be considered significantly better and a ○ indicates a p value greater than 0.95 when comparing the results of ESDT with the other models. A 0 after the name of the method indicates that there is no limit imposed to the tree depth.

Method	D1	D2	D3	D4
ESDT	0.90($\pm$0.10)	0.74($\pm$0.06)	0.74($\pm$0.10)	0.68($\pm$0.07)
DT - gini	0.92($\pm$0.08)	0.70($\pm$0.05) ⋄	0.72($\pm$0.09) ⋄	0.61($\pm$0.08) ⋄
DT - entropy	0.92($\pm$0.07)	0.70($\pm$0.05) ⋄	0.75($\pm$0.07)	0.61($\pm$0.08) ⋄
RF - gini (5)	0.93($\pm$0.08) ○	0.70($\pm$0.05) ⋄	0.79($\pm$0.08) ○	0.57($\pm$0.08) ⋄
RF - gini (10)	0.92($\pm$0.09)	0.70($\pm$0.05) ⋄	0.77($\pm$0.10)	0.57($\pm$0.07) ⋄
RF - entropy (5)	0.93($\pm$0.07) ○	0.70($\pm$0.05) ⋄	0.76($\pm$0.09)	0.57($\pm$0.08) ⋄
RF - entropy (10)	0.92($\pm$0.08)	0.71($\pm$0.05) ⋄	0.79($\pm$0.08) ○	0.57($\pm$0.07) ⋄
Oblique DT	0.91($\pm$0.09)	0.66($\pm$0.05) ⋄	0.70($\pm$0.09) ⋄	0.57($\pm$0.09) ⋄
DT - gini, 0	0.92($\pm$0.08)	0.68($\pm$0.05) ⋄	0.73($\pm$0.09)	0.54($\pm$0.08) ⋄
DT - entropy, 0	0.92($\pm$0.08) ○	0.68($\pm$0.05) ⋄	0.73($\pm$0.08)	0.55($\pm$0.07) ⋄
RF - gini (5), 0	0.93($\pm$0.07) ○	0.69($\pm$0.04) ⋄	0.75($\pm$0.09)	0.56($\pm$0.07) ⋄
RF - gini (10), 0	0.92($\pm$0.09)	0.69($\pm$0.05) ⋄	0.80($\pm$0.09)	0.56($\pm$0.06) ⋄
RF - entropy (5), 0	0.92($\pm$0.10)	0.69($\pm$0.06) ⋄	0.76($\pm$0.09)	0.56($\pm$0.08) ⋄
RF - entropy (10), 0	0.93($\pm$0.08) ○	0.70($\pm$0.05) ⋄	0.78($\pm$0.09) ○	0.57($\pm$0.08) ⋄

βCMA-ES versus CMA-ES. In order to evaluate the effect of the selection mechanism used by βCMA-ES we compare results reported on the synthetic data-sets with those obtained by using CMA-ES, minimizing the objective function $\nu()$ without any other modification. A paired *ttest* comparing AUC values rejected the null hypothesis that mean differences are less than or equal to 0 with

Table 3. Same as Table 2, max depth 10. Mean and standard deviation of AUC for all models over all test folds; a ◇ indicates a p value smaller that 0.05, i.e. ESDT results can be considered significantly better and a ○ indicates a p value greater than 0.95. A 0 after the name of the method indicates that there is no limit imposed to the tree depth.

Method	D1	D2	D3	D4
ESDT	0.92(±0.11)	0.69(±0.07)	0.74(±0.10)	0.67(±0.10) ◇
DT - gini	0.92(±0.09)	0.68(±0.05) ◇	0.72(±0.10)	0.54(±0.08) ◇
DT - entropy	0.92(±0.08)	0.68(±0.05) ◇	0.74(±0.07)	0.57(±0.09) ◇
RF - gini (5)	0.93(±0.08)	0.70(±0.06)	0.75(±0.09)	0.56(±0.09) ◇
RF - gini (10)	0.92(±0.09)	0.71(±0.05)	0.79(±0.10) ○	0.56(±0.08) ◇
RF - entropy (5)	0.93(±0.08)	0.70(±0.05)	0.78(±0.09) ○	0.58(±0.08) ◇
RF - entropy (10)	0.92(±0.07)	0.71(±0.05) ○	0.79(±0.08) ○	0.56(±0.07) ◇
Oblique DT	0.91(±0.09)	0.66(±0.05) ◇	0.70(±0.09) ◇	0.57(±0.09) ◇
DT - gini, 0	0.92(±0.08)	0.68(±0.05) ◇	0.73(±0.09)	0.54(±0.08) ◇
DT - entropy, 0	0.92(±0.08)	0.68(±0.05) ◇	0.73(±0.08)	0.55(±0.07) ◇
RF - gini (5), 0	0.93(±0.07)	0.69(±0.04)	0.75(±0.09)	0.56(±0.07) ◇
RF - gini (10), 0	0.92(±0.09)	0.69(±0.05)	0.80(±0.09) ○	0.56(±0.06) ◇
RF - entropy (5), 0	0.92(±0.10)	0.69(±0.06)	0.76(±0.09)	0.56(±0.08) ◇
RF - entropy (10), 0	0.93(±0.08)	0.70(±0.05)	0.78(±0.09) ○	0.57(±0.08) ◇

$p < 0.0001$, indicating that it can be considered that the use of the selection mechanism improves results.

5   Conclusions and Further Work

The use of the Nash equilibrium concept as a possible tool for splitting data at the node level in decision trees for the binary classification problem is explored. A two-player game in which each sub-node attempts to find the hyperplane that minimizes its entropy is designed, and data is split based on the equilibrium strategy of the game. The Nash equilibrium is approximated by minimizing an objective function that simulates a number of unilateral deviations. The function is constructed so that - given enough deviations - it has a minimum value of 0 at Nash equilibria. CMA-ES, endowed with a new selection scheme, is used to minimize it.

While there is no guarantee that the Nash equilibrium of the game exists or that it is computed by βCMA-ES, it is reasonable to assume that the solutions found present some equilibrium properties making them worth exploring. Numerical results reported by ESDT compared with other tree-based methods indicate the potential of the approach. In order to isolate the effect of using the equilibrium concept, at this point, ESDT is based only on the Nash equilibrium split using all features in the data at each node, and probit classification for predictions, without any feature management mechanism. Further work consists of adding some feature selection mechanisms at the node level to improve scalability. Many other possible modifications and improvements can be used in order to refine classification results. However, there is reasonable evidence to support the assumption that using the Nash equilibrium as a solution concept may benefit the induction of decision trees for the binary classification problem.

Acknowledgements. This work was supported by a grant of the Romanian Ministry of Education and Research, CNCS - UEFISCDI, project number PN-III-P4-ID-PCE-2020-2360, within PNCDI III.

References

1. Banos, R.C., Jaskowiak, P.A., Cerri, R., de Carvalho, A.C.P.L.F.: A framework for bottom-up induction of oblique decision trees. Neurocomputing **135**(SI), 3–12 (2014). https://doi.org/10.1016/j.neucom.2013.01.067
2. Breiman, L., Friedman, J.H., Olshen, R.A., Stone, C.J.: Classification and Regression Trees. Wadsworth and Brooks, Monterey, CA (1984)
3. Breiman, L.: Random forests. Mach. Learn. **45**(1), 5–32 (2001). https://doi.org/10.1023/A:1010933404324
4. Cai, Y., Zhang, H., He, Q., Sun, S.: New classification technique: fuzzy oblique decision tree. Trans. Instit. Measur. Control **41**(8, SI), 2185–2195 (2019). https://doi.org/10.1177/0142331218774614
5. ECNU: oblique decision tree in python. https://github.com/zhenlingcn/scikit-obliquetree (2021)
6. Fawcett, T.: An introduction to ROC analysis. Pattern Recogn. Lett. **27**(8), 861–874 (2006). https://doi.org/10.1016/j.patrec.2005.10.010. rOC Analysis in Pattern Recognition
7. Freedman, D.A.: Statistical Models: Theory and Practice. Cambridge University Press, 2 edn. (2009). https://doi.org/10.1017/CBO9780511815867
8. Hansen, N., Müller, S.D., Koumoutsakos, P.: Reducing the time complexity of the derandomized evolution strategy with covariance matrix adaptation (cma-es). Evol. Comput. **11**(1), 1–18 (2003). https://doi.org/10.1162/106365603321828970
9. Hastie, T., Tibshirani, R., Friedman, J.: The Elements of Statistical Learning. SSS, Springer, New York (2009). https://doi.org/10.1007/978-0-387-84858-7
10. Huysmans, J., Dejaeger, K., Mues, C., Vanthienen, J., Baesens, B.: An empirical evaluation of the comprehensibility of decision table, tree and rule based predictive models. Decis. Support Syst. **51**(1), 141–154 (2011). https://doi.org/10.1016/j.dss.2010.12.003
11. Leroux, A., Boussard, M., Des, R.: Inducing Readable Oblique Decision Trees. In: 2018 IEEE 30th International Conference on Tools With Artificial Intelligence (ICTAI), pp. 401–408 (2018). https://doi.org/10.1109/ICTAI.2018.00069. Volos, Greece, 05-07 Nov 2018
12. Li, Y., Dong, M., Kothari, R.: Classifiability-based omnivariate decision trees. IEEE Trans. Neural Netw. **16**(6), 1547–1560 (2005)
13. McKelvey, R.D., McLennan, A.: Computation of equilibria in finite games. Handbook Comput. Econ. **1**, 87–142 (1996)

14. Murthy, S.K., Kasif, S., Salzberg, S.: A system for induction of oblique decision trees. J. Artif. Intell. Res. **2**, 1–32 (1994)
15. Pedregosa, F., et al.: Scikit-learn: machine learning in Python. J. Mach. Learn. Res. **12**, 2825–2830 (2011)
16. Rosset, S.: Model selection via the AUC. In: Proceedings of the Twenty-First International Conference on Machine Learning, p. 89. ICML 2004, Association for Computing Machinery, New York, NY, USA (2004). https://doi.org/10.1145/1015330.1015400
17. Scholz, M., Wimmer, T.: A comparison of classification methods across different data complexity scenarios and datasets. Expert Syst. Appl. **168**, 114217 (2021). https://doi.org/10.1016/j.eswa.2020.114217
18. Setiono, R., Liu, H.: A connectionist approach to generating oblique decision trees. IEEE Trans. Syst. Man Cybern. Part B Cybern. **29**(3), 440–444 (1999). https://doi.org/10.1109/3477.764880
19. yan Song, Y., Lu, Y.: Decision tree methods: applications for classification and prediction. Shanghai Archiv. Psychiatry **27**, 130–135 (2015)
20. Wickramarachchi, D., Robertson, B., Reale, M., Price, C., Brown, J.: HHCART: an oblique decision tree. Comput. Statist. Data Anal. **96**, 12–23 (2016). https://doi.org/10.1016/j.csda.2015.11.006
21. Wu, X., et al.: Top 10 algorithms in data mining. Knowl. Inf. Syst. **14**(1), 1–37 (2008). https://doi.org/10.1007/s10115-007-0114-2
22. Zaki, M.J., Meira, W.: Data Mining and Machine Learning: Fundamental Concepts and Algorithms. Cambridge University Press, 2 edn. (2020). https://doi.org/10.1017/9781108564175

Evaluating a New Genetic Algorithm for Automated Machine Learning in Positive-Unlabelled Learning

Jack D. Saunders$^{(\boxtimes)}$ ⓘ and Alex A. Freitas ⓘ

School of Computing, University of Kent, Canterbury CT2 7NF, UK
{jds39,A.A.Freitas}@kent.ac.uk

Abstract. Positive-Unlabelled (PU) learning is a growing area of machine learning that aims to learn classifiers from data consisting of a set of labelled positive instances and a set of unlabelled instances, where the latter can be either positive or negative instances, but their label is unknown. There are many PU-learning algorithms, so an exhaustive search to find the best algorithm for a given dataset is computationally unfeasible. We recently proposed GA-Auto-PU, the first Genetic Algorithm-based Automated Machine Learning system for PU learning, and reported its preliminary results. This work presents an improved version of this system with an extended search space to include spy-based techniques, and provides an extensive evaluation of the new and previous versions of this system.

Keywords: Positive-Unlabelled Learning · Classification · Automated Machine Learning (Auto-ML) · Genetic Algorithm

1 Introduction

Positive-Unlabelled (PU) learning is a field of machine learning that aims to learn classifiers from positive and unlabelled data [1]. This differs from binary classification due to the absence of a distinct and accurate negative set. The two instance sets present in PU learning are the labelled positive set, consisting of positive instances that have been labelled as positive, and the unlabelled set, consisting of instances which can be in reality positive or negative, but whose label is unknown. Consequently, the challenge lies in training a classifier that can estimate the likelihood of an instance belonging to the positive class, despite the mixture of negative and positive instances in the unlabelled set. When learning such a classifier, a PU learning algorithm is 'aware' of the difference between the concepts of unlabelled and negative instances. This allows a PU learning algorithm to try to infer which unlabelled instances are negative and which unlabelled instances are positive. In contrast, a traditional classification algorithm applied to PU data lacks such awareness, treating unlabelled instances as a single category alongside the labelled positive class, resulting in a learned classifier that predicts the probability of an instance being labelled [2].

P. Legrand et al. (Eds.): EA 2022, LNCS 14091, pp. 42–57, 2023.
https://doi.org/10.1007/978-3-031-42616-2_4

PU learning naturally occurs in many application domains, such as bioinformatics [2], text-mining [3], pharmacology [4], etc. [1], due to the impracticality of obtaining fully labelled data. E.g., consider datasets where the class variable represents the presence or absence of a disease. Instances with the positive class label (disease) are in general reliable, indicating that the patient was diagnosed earlier. However, the data often also includes patients who did not undergo detailed medical examination yet, essentially 'lack of evidence for the positive class', instead of 'evidence for the negative class' (not having the disease). Hence, if those patients are labelled with the negative class (as it is usually the case), this leads to unreliable negative-class labels. PU learning avoids this by treating all those patients as 'unlabelled', to reflect their unreliability.

Many PU learning algorithms have been proposed (see [1] for a comprehensive review). Hence, finding the optimal PU learning algorithm for specific classification tasks can prove unfeasibly time consuming and computationally expensive, should one use an exhaustive search. Furthermore, algorithm predictive performance is largely dependent on the input data [1]. Therefore, the PU learning area could benefit greatly from an Automated Machine Learning (Auto-ML) system, which selects the best algorithm for a given input dataset, among a pre-defined set of candidate algorithms [5,6].

Our previous short paper [7] recently proposed GA-Auto-PU, the first Genetic Algorithm (GA)-based Auto-ML system for PU learning, and reported its preliminary results. This work presents an improved version of this system with an extended search space and provides a much more extensive evaluation of the new and previous versions of this system, comparing their predictive performance to two popular PU learning methods on three distributions of PU data across 20 datasets, i.e., 60 PU learning problems in total; whilst only 20 PU learning problems were used in [7].

2   Background

2.1   Positive-Unlabelled (PU) Learning

PU learning is a field of machine learning that aims to learn models from datasets that consist of only positive-class and unlabelled instances [1]. PU learning shares the goal of binary classification - accurately predicting the class of an unseen instance by learning to distinguish between two classes. However, as a standard binary classifier requires a training set with two class labels, such a classifier built using PU data would have to treat all unlabelled instances as a separate class, and so will predict the probability of an instance being labelled ($\Pr(s=1)$) instead of the probability of an instance belonging to the positive class ($\Pr(y=1)$) [2] - where s is a variable taking 1 or 0 to indicate whether or not an instance is labelled, and y is the true label of an instance, taking values 1 or 0 to denote the positive or negative class, respectively. In contrast, PU learning models are trained to predict $\Pr(y=1)$ using PU data and have been shown theoretically to improve upon standard binary classifiers when applied to PU datasets [8].

Within the PU learning literature, it is commonly assumed (implicitly or explicitly) that the data adheres to the selected completely at random assumption [2], stating that for the given data, $\Pr(s=1) = \Pr(s=1|x)$, where $\Pr(s=1)$ is the probability of an instance being labelled, and x represents a feature vector. I.e., the labelled examples are selected from the positive distribution irrespective of their features and the labelled set is an independent and identically distributed sample from the positive distribution. There are several approaches to PU learning, including the two-step approach, biased learning, and incorporation of the class prior [1]. The two-step approach is by far the most popular and is the focus of our proposed Auto-ML system.

The first step of this approach identifies a set of reliable negative (RN) instances among the unlabelled set; i.e., a set of instances substantially different from the labelled positive instances and likely not unlabelled positives. The second step builds a classifier to distinguish the labelled positives from the RN set. These two steps use only the training set [9]. Provided that the RN set is an accurate representation of the negative class, this model will predict $\Pr(y=1)$ rather than $\Pr(s=1)$.

This approach makes two assumptions [1]: (a) data separability, assuming a natural division between positive and negative classes; and (b) data smoothness, assuming that similar instances have similar probabilities of belonging to the positive class.

An example of a two-step technique is the "Deep Forest PU" (DF-PU) method [10]. It trains a state-of-the-art deep forest classifier on a random sample of 20% of the unlabelled instances (treated as negative) and all positive instances. The bottom 1% of instances with the lowest probability of belonging to the positive class are added to the RN set. This process is repeated 5 times. A classifier is then trained to distinguish the RN set and the positive set. Our system is compared against DF-PU as a recent state-of-the-art PU learning method.

We also compare our system against a well-known method: S-EM [9]. S-EM uses a spy-based approach, hiding some of the labelled positive instances in the unlabelled set and using them to determine the threshold under which an instance is considered RN. It uses the Expectation Maximisation algorithm [11] and determines the RN threshold as the predicted probability of being positive under which 1% of spy instances fall. Whilst the literature generally refers to two individual steps for two-step methods, this work uses slightly different terminology. We refer to the steps as phases and recognise that "Step 1" often consists of two distinct phases. Hence, when discussing two-step methods, this work references Phase I-A, which extracts an initial RN set; Phase I-B, an optional step using the initial RN set to further extract RN instances from the unlabelled set; and Phase II, "Step 2" in the usual description, which builds a classifier using the positive and RN sets. This notation is advantageous as it recognises that "Step 1" often consists of two distinct phases, and the use of "phase" rather than "step" allows us to reference the individual steps of each phase without confusion.

2.2   Automated Machine Learning (Auto-ML)

Auto-ML is an emerging field of machine learning (ML) that looks to limit the
human involvement in ML applications [5], reducing the demand for domain
experts and allowing those without extensive ML knowledge to operate complex ML pipelines [6]. As algorithm performance is largely dependent on input
data [12], Auto-ML is a useful tool as it searches for the best algorithm specific
to the target ML task.

Reference [13] proposed the Tree-based Pipeline Optimisation Tool (TPOT),
an Auto-ML system using genetic programming (GP). The GP uses tree-based
encoding such that the individuals in the population are ML pipelines. The
functions are pipeline operators and hyperparameters, e.g., specifying the number of trees in a random forest or the number of features selected by filter feature selection methods. The original version of TPOT uses a multi-objective
optimization approach, where each individual is evaluated by both the classification accuracy and the complexity of the pipeline produced, based on the
Non-dominated Sorting Genetic Algorithm II [14], drawing on the well-known
concept of Pareto dominance [15,16]. A drawback of the original version is that
it can produce individuals that represent invalid pipelines, with a large computational cost in terms of evaluation [17]. This issue has been ad-dressed by other
EA-based Auto-ML systems, such as the Resilient Classification Pipeline Evolution system (RECIPE). Like TPOT, RECIPE, proposed by [17], is a genetic
programming system that evolves ML pipelines. However, RECIPE uses a grammar to ensure that all generated individuals are valid, so that it does not waste
resources on invalid individuals. Furthermore, RECIPE evaluates a larger search
space than TPOT which, whilst making for a more complex search space, allows
for a greater variety of solutions [17].

The systems described in this section are for standard binary classification
and are not suitable for PU learning. Hence, we have not compared our system
against any of these. Instead, we have compared against the two PU methods
outlined in Sect. 2.1.

3   The GA-Auto-PU System

The next three Subsections define individual representation and genetic operators used by the GA. These subsections are based on the description of GA-
Auto-PU in [7], with two main differences. First, the current GA version uses
an extended search space of PU learning methods which includes two spy-based
methods, not used in [7]. This involves an extended individual representation and
a new procedure for handling instances used as "spies" (Algorithm 3, described
later). Second, in [7] the GA used a lexicographic multi-objective fitness function for optimising F-measure with higher priority and recall as lower priority
(these measures are defined below). In contrast, in this current paper the GA
uses a simpler fitness function, optimising the F-measure only. This simplifica-
tion was made because further experiments (performed after the writing of [7])
have shown that the simpler fitness function (F-measure only) produces results

qualitatively similar to the more complex lexicographic fitness function, hence the former is now used in this work. The F-measure is defined as follows:

$$\text{F-Measure} = 2 \times \frac{\text{Precision} \times \text{Recall}}{\text{Precision} + \text{Recall}} \quad \text{Precision} = \frac{\text{TP}}{\text{TP} + \text{FP}} \quad \text{Recall} = \frac{\text{TP}}{\text{TP} + \text{FN}}$$

TP is true positive count, FP is false positive count, and FN is false negative count.

3.1 Individual Representation

An individual of the GA represents a candidate solution, i.e., a two-step PU learning method and its hyperparameter configuration. In both versions of the system, each two-step technique is composed of three components: phase I-A, phase I-B, and phase II. Phase I-A of the original system [7] consists of three parameters: $iteration\_count\_1A$, $threshold\_1A$, $classifier\_1A$. The new proposed version of the system introduces three new parameters for phase I-A: spy_flag, spy_rate, and $spy\_tolerance$.

The iteration count determines the number of subsets to split the unlabelled set into when learning a classifier to distinguish between the positive and the unlabelled set. E.g., with an iteration count of 5, each subset contains 20% of the unlabelled data. This helps to handle the imbalance present in many PU learning datasets. The threshold determines the predicted probability required for an instance to be considered a reliable negative (RN) instance. In the literature, the iteration count and threshold are either set heuristically [9] or arbitrarily [10]. The previous version of the system did not generate PU learning methods with heuristic initialisation, but this has been added with the inclusion of spy-based methods. The classifier is the classifier used to predict the RN instances. Spy_flag is a Boolean value used to indicate whether or not to use a spy-based method in Phase I-A. Spy_rate determines the percentage of positive instances to use as spies. $Spy\_tolerance$ determines what percentage of spies can remain in the unlabelled set when the threshold is calculated.

Phase I-B consist of three parameters: $threshold\_1B$, $classifier\_1B$, and $phase\_1B\_flag$. The threshold and the classifier are analogous to those used in phase I-A. The $phase\_1B\_flag$ parameter indicates whether to skip phase I-B or not. Phase I-B is not always utilised in PU learning techniques, and therefore the system will also generate individuals that are able to skip this phase. Phase II simply consists of a single parameter: $classifier\_2$. This classifier will be trained to distinguish the positive set and the RN set. The list of the 10 genes encoded into an individual is shown in Fig. 1.

iteration_count_1A	thresh._1A	classifier_1A	spy_flag	spy_rate	spy_tol.	thresh._1B	classifier_1B	flag_1B	classifier_1B

Fig. 1. Individual representation of the GA.

The genes *Classifier_1A*, *Classifier_1B*, and *Classifier_2* can take the same set of values, representing 18 different candidate classification algorithms: Gaussian naïve Bayes, Random forest, Decision tree, Multilayer perceptron, Support vector machine, Stochastic gradient descent classifier, Logistic regression, K-nearest neighbour, Deep forest, AdaBoost, Gradient boosting classifier, Linear discriminant analysis, Extra tree classifier, Extra trees classifier (an ensemble of Extra trees), Bagging classifier, Bernoulli naïve Bayes, Gaussian process classifier, and Histogram-based gradient boosting classification tree. These values are henceforth referred to as "*Candidate_classifiers*". The candidate values of each gene in the individual representation (defining the search space of PU learning algorithms and their configuration) are as follows:

- *Iteration_count_1A*: { 1, 2, 3, 4, 5, 6, 7, 8, 9, 10 }
- *Threshold_1A*: { 0.05, 0.1, 0.15, 0.2, 0.25, 0.3, 0.35, 0.4, 0.45, 0.5 }
- *Classifier_1A*: { *Candidate_classifiers* }
- *Spy_flag*: { True, False }
- *Spy_rate*: { 0.05, 0.1, 0.15, 0.2, 0.25, 0.3, 0.35 }
- *Spy_tolerance*: { 0, 0.01, 0.02, 0.03, 0.04, 0.05, 0.06, 0.07, 0.08, 0.09, 0.1 }
- *Threshold_1B*: { 0.05, 0.1, 0.15, 0.2, 0.25, 0.3, 0.35, 0.4, 0.45, 0.5 }
- *Classifier_1B*: { *Candidate_classifiers* }
- *Phase_1B_flag*: { True, False }
- *Classifier_2*: { *Candidate_classifiers* }

The size of the search space of the GA is thus calculated as follows:
$10 \times 10 \times 18 \times 2 \times 7 \times 11 \times 10 \times 18 \times 2 \times 18 = 1,796,256,000$ possible candidate solutions.

PU learning solutions that do not find any RN instances get a fitness of 0.

3.2   Outline of the Underlying Genetic Algorithm (GA)

Algorithm 1 outlines the procedure that the GA follows to evolve a PU learning algorithm. Initially, a Population of *Population_size* individuals is generated (step 1). The configuration (genome) of the individual is checked against a list of previously assessed configurations, and if it has not already been assessed, the Fitness of *Individual* is calculated (steps 3–4) as described in Sect. 3.3, including Algorithms 2, 3, 4 and 5. If the configuration has already been assessed, the fitness values of the previous assessment are assigned to the individual (steps 5–6). Once all individuals have been evaluated, the fittest *Individual* is saved for the following generation (step 7). A new population is created from Population after undergoing tournament selection (step 8), and *New_pop* then undergoes crossover (step 9) and mutation (step 10). Finally, Population is set as *New_pop* and *Fittest_individual* (step 11). This process of fitness calculation, selection, crossover, mutation, and elitism is repeated #generations times. The fitness of an individual is assigned as the F-measure achieved over the 5 folds of the internal cross-validation (applied to the training set).

Both system versions employ uniform crossover and mutation as search operators, randomly replacing gene values with candidate values.

Algorithm 1. Outline of the GA procedure

1: *Population* = Generate population();
2: **for** *#generations* times **do**
3:     **if** *Individual* configuration has not already been assessed **then**
4:         Assess fitness(*Individual*, *Training set*); // see Algorithms 2, 3, 4, 5
5:     **else**
6:         *IndividualFitness* value is set as the output of the previous assessment;
7:     *Fittest_individual* = Get fittest individual(*Population*);
8:     *New_pop* = 100 individuals selected from *Population* by tournament selection;
9:     Individuals in *New_pop* undergo crossover;
10:     Individuals in *New_pop* undergo mutation;
11:     *Population* = *New_pop* ∪ {*Fittest_individual*};

3.3   Fitness Evaluation

Recall that each individual encodes three classification algorithms, which are used in phases I-A, I-B and II of the PU learning system. Fitness evaluation involves applying these algorithms (with the possible exception of the algorithm for the optional Phase I-B) to the training set. To describe the fitness evaluation process, we use this notation:

RN: The set of reliable negative instances.

P: The set of labelled positive instances.

U: The set of unlabelled instances.

$P+RN$: The combined set of labelled positive and reliable negative instances.

$P(y = 1)$: The probability of an instance being positive, as calculated by the classifier.

Algorithm 2. Assess Fitness (*Individual*, *Trainingset*)

1: Split *Trainingset* into 5 *Learning* and *Validation sets*;
2: **for** each *Learningset* and corresponding *Validationset* **do**
3:     P = all labelled positive instances in *Learningset*;
4:     U = all unlabelled instances in *Learningset*;
5:     **if** *Spy_flag* **then**
6:         RN, U = Phase I-A-Spies(P, U) // call Algorithm 3
7:     **else**
8:         RN, U = Phase I-A(P, U); // call Algorithm 4
9:     **if** *Phase_1B_flag* **then**
10:         RN, U = Phase I-B($P + RN, U$); // call Algorithm 5
11:     Train *Classifier_2* to distinguish P and RN;
12:     Classify *Validationset*;
13: *Individual's Fitness Value* = F-measure over the 5 *Validation sets*;

The fitness of each individual is computed as specified in Algorithm 2. The Training set is split into 5 folds for internal cross-validation, creating 5 pairs of

---

Algorithm 3. Phase I-A-Spies(P, U)

1: $RN = \{\ \}$;
2: $Sets = $ Split U into $Iteration\_count\_1A$ subsets;
3: **for** every Set in $Sets$ **do**
4:      $Spies = spy\_rate\%$ instances, randomly selected from P;
5:      $Spy\_set = Set \cup Spies$;
6:      Run EM($Classifier\_1A$, P, Spy_set);
7:      Classify all instances in Spy_set;
8:      Set $threshold$ to a value such that $spy\_tolerance\%$ spies have $P(y = 1)$ less
        than $threshold$;
9:      **for** each unlabelled $Instance$ in Spy_set **do**
10:         **if** $P(y = 1) < threshold$ **then**
11:             $RN = RN \cup Instance$, $U = U - Instance$
12: **Return** RN, U;

Learning and Validation sets (step 1). For each pair of Learning and Validation sets, all labelled positive instances are added to P (step 3) and all unlabelled instances are added to U (step 4). The RN set is determined with either the phase I-A-Spies(P, U) or phase I-A(P, U) algorithm, depending on the spy_flag parameter, which returns a refined U set (steps 5–8, executing Algorithm 3 or 4). If the flag for running phase I-B is set to true, RN and U sets are further defined with the phase I-B($P + RN, U$) algorithm (steps 9–10, executing Algorithm 5). $Classifier\_2$ is then trained to distinguish P and RN (step 11), and then used to classify the Validation set (step 12). The Fitness Value of the Individual is assigned as the F-measure over the 5 Validation set classifications (step 13).

Algorithm 3 outlines Phase I-A of the two-phase PU learning method when spy_flag is True. The RN set is initialised empty (step 1). The set U of unlabelled instances is split into $Iteration\_count\_1A$ subsets (step 2). For each Set in the list of subsets, Spies is initialised with $spy\_rate\%$ of instances of P, randomly selected (step 4) and Set and Spies are combined to form Spy_set (step 5). Next, the Expectation Maximisation (EM) algorithm is run [9], tuning $Classifier\_1A$ on P and Spy_set (step 6). All instances in Spy_set are classified and the threshold is set so that $spy\_tolerance\%$ of spies have $P(y = 1)$ less than threshold (steps 7–8). For each unlabelled Instance in Spy_set (excluding the spies), if $P(y = 1)$ is less than threshold, they are added to RN and removed from U (steps 9–11). The resulting RN and U sets are then returned.

Phase I-A of the two-phase PU learning method when Spy_flag is False is described in Algorithm 4. The RN set is initialised as an empty set (step 1). The set U of unlabelled instances is split into $Iteration\_count\_1A$ subsets (step 2). For each Set in the list of subsets, $Classifier\_1A$ is trained to distinguish P and Set (step 4) and used to classify all unlabelled instances in Set (instances previously treated as the negative set during training) (step 5). For each unlabelled Instance, if the instance's calculated $P(y = 1)$ is less than $Threshold\_1A$ then Instance is added to RN and removed from U (step 8). The resulting RN and U sets are then returned.

Phase I-B of the two-phase learning method is described in Algorithm 5. $Classifier\_1B$ is trained to distinguish the positive and reliable negative instances in $P + RN$ (step 1) and the resulting classifier is then used to classify U (step 2). For each Instance in U, if the Instance's calculated $P(y = 1)$ is less than $Threshold\_1B$, Instance is added to RN and removed from U (step 5). The resulting RN and U sets are returned (step 6).

Algorithm 4. Phase I-A(P, U)

1: $RN = \{\ \}$;
2: $Sets = $ Split U into $Iteration\_count\_1A$ subsets;
3: **for** every Set in $Sets$ **do**
4:     Train $Classifier\_1A$ on P and Set;
5:     Classify all unlabelled instances in Set;
6:     **for** each unlabelled $Instance$ in Set **do**
7:         **if** $P(y = 1) < Threshold\_1A$ **then**
8:             $RN = RN \cup Instance, U = U - Instance$
9: **Return** RN, U;

Algorithm 5. Phase I-B$(P + RN, U)$

1: Train $Classifier\_1B$ on $P + RN$;
2: Classify U;
3: **for** each $Instance$ in U **do**
4:     **if** $P(y = 1) < Threshold\_1B$ **then**
5:         $RN = RN \cup Instance, U = U - Instance$
6: **Return** RN, U;

---

4   Datasets and Experimental Methodolody

The experiments used a stratified 5-fold cross-validation procedure. Each dataset is split into 5 folds of about the same size with about the same class distribution, and then the methods are run 5 times. Each time, a different fold is used as the test set, and the other 4 folds are used as the training set. For each training set, we run the Auto-ML systems to evolve the best individual configuration. Then, a PU learning classifier is built from the training set with the configuration defined by that best individual. The classifier is then used to predict all instances in the test set. Precision, recall, and F-measure are calculated. This process is repeated for the 5 pairs of training and test sets in the 5-fold cross-validation, and the reported results are the average over the 5 test set results. Each method is evaluated using the same procedure, with the same 5 folds.

We compare the predictive performance of the two versions of the Auto-ML system, GA-Auto-PU [7] (called GA-1) and GA-Auto-PU with the extended search space (GA-2), and two well-established PU learning methods: DF-PU [10] and S-EM [9]. Performance is measured mainly by the F-measure, the most used measure in PU learning [18], but we also report a summary of precision and recall results, for completeness.

The two versions of the GA used the following hyperparameter settings. *#Generations* (number of generations) set to 50. *Population_size* (number of individuals in the population) set to 101 (100 individuals generated using crossover, 1 saved with elitism). *Crossover_prob*, the probability that two individuals will undergo crossover, set to 0.9. *Gene_crossover_prob*, the probability that a specific gene will be swapped when the individuals undergo uniform crossover, set to 0.5. *Mutation_prob*, the probability that an individual's gene will undergo mutation, set to 0.1. *Tournament_size* set to 2. Code for both versions of the GA can be found at https://github.com/jds39/GA-Auto-PU.

Regarding statistical analysis, for each performance measure (F-measure, recall, and precision), we compare the performance of the new GA-2 against GA-1 and the above two PU learning methods using the non-parametric Wilcoxon Signed-Rank test [19]. As this involves testing multiple hypotheses per measure, we use the well-known Holm correction [20] for multiple hypotheses. This procedure involves ranking the p-values from the smallest to largest (i.e., from most to least significant), and adjusting the significance level α according to the p-values' ranking. We set $\alpha=0.05$ as usual, so the first p-value (p_1) is statistically

Table 1. Dataset characteristics.

Dataset	No. instances	No. features	Positive-class %
Alzheimer's [21]	354	9	10.73%
Autism [22]	288	15	48.26%
Breast cancer Coimbra [22]	116	9	55.17%
Breast cancer Wisconsin [22]	569	30	37.26%
Breast cancer mutations [23]	1416	53	32.42%
Cervical cancer [22]	668	30	2.54%
Cirrhosis [24]	277	17	25.72%
Dermatology [22]	359	34	13.41%
Pima Indians Diabetes [22]	769	8	34.90%
Early Stage Diabetes [25]	521	17	61.54%
Heart Disease [22]	304	13	54.46%
Heart Failure [26]	300	12	32.11%
Hepatitis C [22]	590	13	9.51%
Kidney Disease [22]	159	24	27.22%
Liver Disease [22]	580	11	71.50%
Maternal Risk [22]	1014	6	26.82%
Parkinsons [22]	196	22	75.38%
Parkinsons Biomarkers [27]	131	29	23.08%
Spine [22]	311	6	48.39%
Stroke [28]	3427	15	5.25%

significant if $p_1 < 0.017$. If this condition is not satisfied, the procedure stops and p_1, p_2 and p_3 are deemed non-significant. If p_1 is deemed significant, we then evaluate p_2, which is deemed significant if $p_2 < 0.025$. If p_2 is deemed significant, we then evaluate p_3, which is deemed significant if $p_3 < 0.05$.

Table 1 shows the characteristics of each dataset. These datasets are binary datasets that are engineered to PU datasets by hiding $\delta\%$ of the positive instances in the negative set, thus creating an unlabelled set - a common practice in PU learning [18]. Note, the Positive-class % column shows the percentage of positive instances before some are hidden in the unlabelled set. δ takes the values 20%, 40%, and 60%, as it is important to test on varying distributions [18].

5   Computational Results

Table 2 shows the F-measure values achieved by all methods across the datasets for $\delta = 20\%$. GA-2 achieved the highest number of wins with 10.5, followed by GA-1 with 7.5. GA-2 outperformed DF-PU and S-EM with statistical significance (p = 0.00002, Holm's adjusted $\alpha = 0.017$ for DF-PU; p = 0.0008, Holm's

Table 2. F-measure values achieved by the four methods for $\delta = 20\%$.

Dataset	GA-1	GA-2	DF-PU	S-EM
Alzheimer's	0.529	**0.548**	0.195	0.321
Autism	0.960	**0.982**	0.648	0.820
Breast cancer Coi	0.705	**0.711**	0.697	**0.711**
Breast cancer Wis	0.954	**0.956**	0.543	0.898
Breast cancer mut.	0.893	**0.896**	0.489	0.892
Cervical cancer	0.828	**0.867**	0.061	0.054
Cirrhosis	**0.573**	0.446	0.405	0.436
Dermatology	0.860	**0.901**	0.228	0.718
Pima I. Diabetes	**0.677**	0.642	0.516	0.534
Early Diabetes	0.958	**0.978**	0.761	0.792
Heart Disease	**0.843**	0.836	0.705	0.811
Heart Failure	**0.770**	0.751	0.487	0.529
Hepatitis C	**0.953**	0.944	0.176	0.695
Kidney disease	0.976	0.925	0.428	**1.000**
Liver disease	**0.834**	0.831	**0.834**	0.816
Maternal health	0.476	**0.862**	0.403	0.454
Parkinson's	0.860	**0.935**	0.856	0.815
Parkinson's Biom.	**0.476**	0.282	0.354	0.333
Spine	0.652	**0.923**	0.652	0.820
Stroke	**0.474**	0.241	0.086	0.102
Total wins	7.5	**10.5**	0.5	1.5

adjusted $\alpha = 0.025$ for S-EM), but there was no significant difference between GA-2 and GA-1 (p $= 0.7841$, $\alpha = 0.05$).

This trend continued for $\delta = 40\%$, as shown by Table 3. GA-2 achieved the highest number of wins with 8, followed by GA-1 with 5. GA-2 significantly outperformed DF-PU (p $= 0.0003$, adjusted $\alpha = 0.017$) and S-EM (p $= 0.0073$, adjusted $\alpha = 0.025$), but there was no significant difference between GA-2 and GA-1 (p $= 0.7562$, $\alpha = 0.05$).

For $\delta = 60\%$, in Table 4, GA-1 and S-EM performed best with 6 wins each, followed by GA-2 with 5 and DF-PU with 3. GA-2 significantly outperformed DF-PU (p $= 0.0023$, adjusted $\alpha = 0.017$), but there was no significant difference between GA-2 and GA-1 or S-EM (p $= 0.4980$, adjusted $\alpha = 0.025$ for GA-1; p $= 0.5706$, $\alpha = 0.05$, for S-EM). In total, across all Tables, GA-2 performed best with 23.5 wins, followed by GA-1 with 18.5, S-EM with 10.5, DF-PU with 7.5.

Table 3. F-measure values achieved by the four methods for $\delta = 40\%$.

Dataset	GA-1	GA-2	DF-PU	S-EM
Alzheimer's	0.551	**0.576**	0.194	0.37
Autism	0.927	**0.94**	0.648	0.841
Breast cancer Coi.	0.687	0.671	**0.711**	0.704
Breast cancer Wis.	0.932	**0.936**	0.543	0.903
Breast cancer mut.	0.868	0.739	0.489	**0.893**
Cervical cancer	**0.903**	0.839	0.042	0.053
Cirrhosis	**0.464**	0.397	0.401	0.442
Dermatology	0.78	**0.896**	0.229	0.718
Pima I. Diabetes	**0.649**	0.646	0.516	0.526
Early Diabetes	**0.895**	0.887	0.756	0.859
Heart Disease	0.801	0.78	0.705	**0.828**
Heart Failure	0.652	**0.67**	0.486	0.508
Hepatitis C	0.771	**0.863**	0.171	0.708
Kidney disease	0.988	0.951	0.428	**1**
Liver disease	0.803	0.817	**0.832**	0.587
Maternal health	0.812	**0.813**	0.395	0.434
Parkinson's	0.836	0.843	**0.86**	0.748
Parkinson's Biom.	0.265	0.259	**0.354**	0.261
Spine	0.907	**0.917**	0.652	0.839
Stroke	**0.255**	0.239	0.094	0.102
Total wins	5	**8**	4	3

Table 5 summarises the Precision and Recall values achieved by each method, showing the number of wins (out of 20 datasets) of each method and whether

GA-2 performed statistically significantly better or worse than another method, for each measure and for each $\delta = 20\%$, 40%, 60% (the full results per dataset are not shown due to lack of space). In terms of recall, DF-PU performed best overall, with GA-2 performing worst. GA-2 performed significantly worse than DF-PU for all 3 δ values. However, DF-PU generally predicted almost all instances as positive, thus achieving near 100% recall, but near 0% precision. Such classification is unhelpful, representing a bad trade-off between precision and recall, which led to the inferior results for DF-PU regarding F-measure, as shown earlier. GA-2 performed best in terms of Precision, significantly outperforming both DF-PU and S-EM for the 3 δ values.

The performance of GA-2 did come at a computational cost, compared with DF-PU and S-EM. Whilst DF-PU and S-EM took about 4.9 min and 1.5 min on average per dataset respectively, GA-2 took about 3.7 h. As such, GA-2 was 45x slower than DF-PU, and 150x slower than S-EM. All experiments were run on a 48 core GPU with 256GB of memory.

Table 4. F-measure values achieved by the four methods for $\delta = 60\%$.

Dataset	GA-1	GA-2	DF-PU	S-EM
Alzheimer's	0.456	**0.529**	0.171	0.373
Autism	0.91	**0.927**	0.645	0.835
Breast cancer Coi.	0.51	0.553	0.697	**0.699**
Breast cancer Wis.	**0.906**	0.866	0.539	0.904
Breast cancer mut.	0.854	0.872	0.485	**0.892**
Cervical cancer	**0.714**	0.35	0.044	0.046
Cirrhosis	0.443	0.204	0.405	**0.459**
Dermatology	**0.828**	0.692	0.219	0.719
Pima I. Diabetes	0.606	**0.634**	0.515	0.544
Early Diabetes	**0.93**	0.894	0.759	0.793
Heart Disease	0.785	0.786	0.702	**0.829**
Heart Failure	**0.674**	0.671	0.482	0.557
Hepatitis C	0.588	**0.61**	0.160	0.609
Kidney disease	0.754	0.806	0.428	**0.951**
Liver disease	0.804	0.748	**0.834**	0.788
Maternal health	0.735	**0.738**	0.390	0.438
Parkinson's	0.818	0.792	**0.860**	0.762
Parkinson's Biom.	0.233	0.28	**0.367**	0.331
Spine	0.818	0.761	0.652	**0.83**
Stroke	**0.255**	0.243	0.094	0.102
Total wins	**6**	5	3	**6**

Table 5. Summary of Precision and Recall results across all datasets.

	Number of wins regarding Recall				Statistically significant results (> means better, < means worse)
δ	GA-1	GA-2	DF-PU	S-EM	
20%	3.83	0	10.33	5.83	GA-2 < DF-PU (p = 0.00001)
40%	0	0	14.5	5.5	GA-2 < DF-PU (p = 0.00002)
					GA-2 < S-EM (p = 0.0136)
60%	0	0	14	6	GA-2 < DF-PU (p = 0.000002)
					GA-2 < S-EM (p = 0.00001)
δ	Number of wins regarding Precision				Statistically significant results
20%	7.83	10.83	0	1.33	GA-2 > DF-PU (p = 0.00001)
					GA-2> S-EM (p = 0.0003)
40%	7.33	11.33	0	1.33	GA-2 > DF-PU (p = 0.001)
					GA-2 > S-EM (p = 0.001)
60%	9.33	10.33	0	0.33	GA-2 > DF-PU (p = 0.00001)
					GA-2 > S-EM (p = 0.0001)

6   Conclusions

We recently proposed GA-Auto-PU, the first GA-based automated machine learning method for PU learning [7]. In this work we presented an improved version of the system which features an extended search space, incorporating spy-based heuristic methods into the genes of the individuals, which allows the creation of more sophisticated PU learning algorithms. This new GA-Auto-PU version was extensively compared against two established and well-performing PU learning methods, as well against the previous version of the system, across three distributions of engineered PU learning data in 20 datasets (representing in total 60 different PU learning problems). The new version of the system out-performed the previous version in general, and the new version outperformed the PU learning baselines with statistical significance in regard to F-measure, the most used performance measure in PU learning [18]. An analysis of the results for recall and precision (used to compute the F-measure) showed that the new system significantly outperformed the two baseline methods regarding precision, but it is significantly outperformed by the two baselines in most cases regarding recall.

Future work will look to explore other search and optimisation methods, such as Bayesian Optimisation, as well as expanding the GA's search space to include other types of PU learning methods.

References

1. Bekker, J., Davis, J.: Learning from positive and unlabeled data: a survey. Mach. Learn. **109**(4), 719–760 (2020)
2. Elkan, C., Noto, K.: Learning classifiers from only positive and unlabeled data. In: Proceedings of the 14th ACM SIGKDD International Conference on Knowledge Discovery and Data Mining, pp. 213–220 (2008)
3. Li, X., Liu, B.: Learning to classify texts using positive and unlabeled data. In: Proceedings of the 18th International Joint Conference on Artificial Intelligence, vol. 3, pp. 587–592 (2003)
4. Zheng, Y., Peng, H., Zhang, X., Zhao, Z., Gao, X., Li, J.: Ddi-pulearn: a positive-unlabeled learning method for large-scale prediction of drug-drug interactions. BMC Bioinform. **20**(19), 1–12 (2019)
5. Q. Yao, et al.: Taking human out of learning applications: a survey on automated machine learning. arXiv preprint arXiv:1810.13306 (2018)
6. He, X., Zhao, K., Chu, X.: AutoML: a survey of the state-of-the-art. Knowl.-Based Syst. **212**, 106622 (2021)
7. Saunders, J.D., Freitas, A.A.: Ga-auto-PU: a genetic algorithm-based automated machine learning system for positive-unlabeled learning. In: Proceedings of the GECCO 2022 Companion (Genetic and Evolutionary Computation Conf.), pp. 288–291. ACM (2022)
8. Niu, G., du Plessis, M., Sakai, T., Ma, Y., Sugiyama, M.: Theoretical comparisons of positive-unlabeled learning against positive-negative learning. In: Proceedings of the 30th International Conference on Neural Information Processing Systems (NIPS 2016), pp. 1207–1215 (2016)
9. Liu, B., Lee, W.S., Yu, P.S., Li, X.: Partially supervised classification of text documents. In: International Conference on Machine Learning, vol. 2, pp. 387–394 (2002)
10. Zeng, X., Zhong, Y., Lin, W., Zou, Q.: Predicting disease-associated circular RNAs using deep forests combined with positive-unlabeled learning methods. Brief. Bioinform. **21**(4), 1425–1436 (2020)
11. Dempster, A., Laird, N.M., Rubin, D.: Maximum likelihood from incomplete data via the EM algorithm. J. Roy. Stat. Soc. B **39**, 1–38 (1977)
12. Brazdil, P., Carrier, C.G., Soares, C., Vilalta, R.: Metalearning: Applications to Data Mining. Springer, Heidelberg (2008). https://doi.org/10.1007/978-3-540-73263-1
13. Olson, R.S., Bartley, N., Urbanowicz, R.J., Moore, J.H.: Evaluation of a tree-based pipeline optimization tool for automating data science. In: Proceedings of the Genetic and Evolutionary Computation Conference (GECCO 2016), pp. 485–492 (2016)
14. Deb, K., Pratap, A., Agarwal, S., Meyarivan, T.A.M.T.: A fast and elitist multi-objective genetic algorithm: NSGA-II. IEEE Trans. Evol. Comput. **6**(2), 182–197 (2002)
15. Deb, K.: Multi-objective Optimization Using Evolutionary Algorithms. Wiley, Hoboken (2001)
16. Freitas, A.A.: A critical review of multi-objective optimization in data mining: a position paper. ACM SIGKDD Explorations Newsl **6**(2), 77–86 (2004)
17. de Sá, A.G., Pinto, W.J.G., Oliveira, L.O.V., Pappa, G.L.: Recipe: a grammar-based framework for automatically evolving classification pipelines. In: Proceedings of the European Conference on Genetic Programming, pp. 246–261 (2017)

18. Saunders, J.D., Freitas, A.A.: Evaluating the predictive performance of positive-unlabelled classifiers: a brief critical review and practical recommendations for improvement. ACM SIGKDD Expl. **24**(2), 5–11 (2022)
19. Wilcoxon, F., Katti, S.K., Wilcox, R.A.: Critical values and probability levels for the Wilcoxon rank sum test and the Wilcoxon signed rank test. Sel. Tables Math. Stat. **1**, 171–259 (1963)
20. Demšar, J.: Statistical comparisons of classifiers over multiple data sets. J. Mach. Learn. Res. **7**, 1–30 (2006)
21. Marcus, D.S., Fotenos, A.F., Csernansky, J.G., Morris, J.C., Buckner, R.L.: Open access series of imaging studies: longitudinal MRI data in nondemented and demented older adults. J. Cogn. Neurosci. **22**(12), 2677–2684 (2010)
22. Asuncion, A., Newman, D.: UCI machine learning repository (2007). http:// archive.ics.uci.edu/ml
23. Pereira, B., et al.: The somatic mutation profiles of 2,433 breast cancers refine their genomic and transcriptomic landscapes. Nat. Commun. **7**(1), 1–16 (2016)
24. Fleming, T.R., Harrington, D.P.: Counting Processes and Survival Analysis. Wiley, New York (1991)
25. Islam, M.F., Ferdousi, R., Rahman, S., Bushra, H.Y.: Likelihood prediction of diabetes at early stage using data mining techniques. In: Computer Vision and Machine Intelligence in Medical Image Analysis, pp. 113–125 (2020)
26. Chicco, D., Jurman, G.: Machine learning can predict survival of patients with heart failure from serum creatinine and ejection fraction alone. BMC Med. Inform. Decis. Mak. **20**(1), 1–16 (2020)
27. Hlavnička, J., Čmejla, R., Tykalová, T., Šonka, K., Růžička, E., Rusz, J.: Automated analysis of connected speech reveals early biomarkers of Parkinson's disease in patients with rapid eye movement sleep behaviour disorder. Sci. Rep. **7**(1), 1–10 (2017)
28. Emon, M.U., Keya, M.S., Meghla, T.I., Rahman, M.M., Al Mamun, M.S., Kaiser, M.S.: Performance analysis of machine learning approaches in stroke prediction. In: 2020 4th International Conference on Electronics, Communication and Aerospace Technology (ICECA), pp. 1464–1469 (2020)

Neural Network-Based Virtual Analog Modeling

Tara Vanhatalo[1,2,3]($\boxtimes$) (iD), Pierrick Legrand[1], Myriam Desainte-Catherine[2],
Pierre Hanna[2](iD), Antoine Brusco[3], Guillaume Pille[3], and Yann Bayle[3](iD)

[1] Inria Bordeaux Sud-Ouest, Institute of Mathematics of Bordeaux,
UMR 5251 CNRS, University of Bordeaux, 33405 Talence, France
tara.vanhatalo@u-bordeaux.fr
[2] University of Bordeaux, CNRS, Bordeaux INP, LaBRI, UMR 5800, 33400 Talence,
France
[3] Orosys, 34980 Saint-Gély-du-Fesc, France

Abstract. Vacuum tube amplifiers present sonic characteristics often
coveted by musicians, that are due to the distinct distortion of their
circuits and accurately modeling such effects can be a challenging task.
A recent rise in machine learning has lead to the ubiquity of neural
networks in all fields including virtual analog modeling. This has lead
to the appearance of a variety of architectures tailored to this task. We
aim to provide an overview of the current state of the research in neural
emulation of distortion circuits.

Keywords: Audio effects modeling · Neural networks · Deep learning

1 Introduction

Guitarists tend to prefer the sound of vacuum tube amplifiers. Yet, their short-
comings include elevated cost and weight, and high power consumption. Digital
emulation can circumvent some downsides and is divided into three categories.
1) White-box methods emulate each electronic component. 2) Black-box tries to
match the output signal by applying custom functions to the input. 3) Gray-box
methods comprise a block-oriented structure inspired by internal information of
the device but disregard the behaviour of each of the individual components.
These "traditional" methods present some drawbacks. White-box methods are
time-consuming as they require hundreds of electronic components to be determined by hand for each amplifier. Moreover with white and gray-box methods it
is necessary to repeat the modeling process for each circuit. Black-box methods
can struggle to accurately approximate the nonlinear mapping of the amplifier.
The computational cost of these methods is also prohibitive for real-time (RT)
use, a fundamental factor to be considered in Virtual Analog (VA) modeling.
Neural networks seem suited for tube amplifier emulation as their nonlinear
activation functions resemble vacuum tubes distortion. They also reduce the
time needed to create a new model as only the training data needs to change.

P. Legrand et al. (Eds.): EA 2022, LNCS 14091, pp. 58–72, 2023.
https://doi.org/10.1007/978-3-031-42616-2_5

This work presents an overview of newly emerging neural approaches in distortion modeling with the aim of identifying future research topics. The structure is as follows: White, gray, and black-box approaches are first introduced. The neural networks used for amplifier emulation are then detailed. Finally, the discussion underlines the future research avenues to improve upon.

2 Traditional Methods

2.1 White-Box

White-box methods model the physical device using internal information of the amplifier to establish a system of differential equations. The main methods are state-space models, Modified Nodal Analysis (MNA), the port-hamiltonian (PH) formalism and Wave Digital Filters (WDF). Nodal analysis is used in circuit simulation techniques to establish the system of equations to be solved in matrix form. MNA extends this to incorporate auxiliary equations into the system [35]. The most well-known electronic circuit simulator, SPICE (Simulation Program with Integrated Circuit Emphasis), uses a combination of component-wise discretization of the circuit and MNA to create the system. The WDF approach constructs digital filters based on the traveling-wave formulation of the physical elements of the device [35]. State-space models rely on the principle that the equations of motion for any physical system may be formulated in terms of the state of the system: $x'(t) = f_t(x(t), u(t))$ where $x(t)$ is the state of the system at time t, $u(t)$ is a vector of external inputs and the function f_t specifies how $x(t)$ and $u(t)$ cause a change in the state. The PH formalism is a state-space representation that is structured based on the various energies of a system and their dynamics. Each of these methods translates the circuit diagram of an amplifier into a set of equations that completely describes it which is then discretized in order to be solved. To establish these equations, access to the circuit diagram or study of the internal structure of the amplifier is necessary and results in a labor-intensive task. Moreover, the resolution of these equations relies on computationally expensive iterative methods or storing lookup tables and the component values measured can also introduce inaccuracies into the emulation.

2.2 Gray-Box

Gray-box methods alleviate some of the labour of white-box as they model the amplifier using conceptual processing blocks (i.e. dynamic, linear, nonlinear), simplifying the model [5]. The Wiener-Hammerstein method has been described as the "fundamental paradigm of electric guitar tone" [10] and is used in commercial products such as Fractal Audio's Axe-Fx [10]. Eichas and Zölzer [5] proposed using iterative methods to optimize the block topology. Gray-box methods require less CPU than white-box methods but struggle with the emulation more.

2.3   Black-Box

Black-box methods only require the input and output signals of an amplifier to emulate it. They can also replicate idiosyncrasies that can exist in the analog devices. The main methods are Volterra series, dynamic convolution, block-oriented structures, and kernel regression. Dynamic convolution is a variant of a method used for linear systems in which instead of a single impulse response used to derive the transfer function of the DUT, multiple impulses are used at different amplitudes to approximate the nonlinear behavior [20]. Gillespie et al. [12] used a Support Vector Machine (SVM) to emulate a common-cathode tube amplifier via kernel regression. While this method is theoretically solid, the choice of the mappings and kernel function can be difficult. Artificial Neural Networks (ANN) are a class of Machine Learning (ML), black-box method that use neurons and nonlinear activation functions frequently found in VA modeling to emulate various nonlinearities. Similarities between a number of ANN and certain black-box methods can be drawn and suggest that they would be well suited to the task of tube amplifier emulation and will be discussed in a future section. Volterra series [27] are functional expansions of multidimensional convolution kernels that emulate the "memory" effect of an amplifier as the output of the nonlinear system is dependent on the input at previous times steps. The block-oriented structures presented previously can also fall under the scope of black-box methods [6]. Overall, these methods remain computationally expensive and/or tend to struggle when simulating very high levels of nonlinearity.

3   Neural Network-Based Methods

ANN are a class of ML algorithm whose basic structure is made up of neurons and often nonlinear activation functions. These neurons are organized into layers and comprise multiplication by weights, with optional bias added, followed by the activation. The network parameters are learnt via minimization of a distance between the target and the network output, enabling the network to learn complex nonlinear mappings making this method suited to distortion modeling. However, amplifier modeling via ANN is not a straight forward task. The deep models used in other fields can pose a problem for RT use. Processing raw waveforms in the time domain means that we have to deal with high temporal dimensionality because of the sampling rates for high-quality audio. This increases the computational cost and the complexity of the task, rendering music Deep Learning (DL) challenging. A number of DL architectures have appeared in distortion circuit modeling in recent years including various configurations of both convolutional and recurrent layers. Here, we study these recent works.

3.1   Architectures

The main categories of architectures that have populated the state-of-the-art are convolutional, recurrent, and hybrid models. The first instances of Convolutional Neural Networks (CNN) for distortion effects modeling were applied to the emulation of amplifiers [3] and for distortion pedals [4], where Damskägg et al. use a feedforward variant of the WaveNet from [23], originally for speech synthesis. This variant contains dilated causal convolutions. A dilated convolution uses increased kernel size to include spaces between elements for larger field of view, in order to better model long-term dependencies, without increased computational cost. These dilated convolutions are causal so only past information is used. Two models of different sizes are presented [3] and compared to a block-oriented model [6] and a Multi-Layer Perceptron (MLP) on modeling a Fender Bassman 56F-A preamplifier. The larger WaveNet outperformed the others in objective and subjective evaluation. Their follow-up article [4] focuses more on the RT possibilities of the WaveNet applied to pedal emulation. Their WaveNet is slightly modified in this work. A trade-off between accuracy of the method and computational load as well as the minimum amount of data required for training was studied. For this, three different configurations were compared on three effects pedals: Ibanez Tube Screamer, Boss DS-1, and an Electro-Harmonix Big Muff Pi. A modified version of the feedforward WaveNet, Temporal Convolutional Network (TCN), was used for more efficient computation for RT use of models with more complex nonlinear behaviour [31]. The authors show that by using shallow networks with large dilation factors comparable performance was achieved with greater efficiency. Both causal and noncausal versions of this modified architecture were tested to emulate a LA-2A dynamic range compressor. They achieved similar performance but the noncausal variants performed slightly better in the time domain. The results of listening tests indicated that a small difference was perceived in the models compared to the reference.

The first article presenting the use of Recurrent Neural Networks (RNN) for vacuum tube amplifier modeling uses a Nonlinear AutoRegressive eXogenous (NARX) network [2]. A NARX network is similar to plain RNN but with limited connectivity to remedy the vanishing and exploding gradient problems frequently encountered. The audio quality of this method was reported to be low when modeling a 4W Vox AC4TV tube amplifier either due to insufficient training or limited model capacity. A follow-up work [36] studies the use of Long Short Term Memory (LSTM), first proposed in [11], for the task of amplifier modeling. Again a 4W Vox AC4TV was chosen. LSTM is a RNN variant that incorporates the use of forget, input and output gates to control information flow through each recurrent layer in order to again avoid gradient problems of regular RNN. The models used were configured as multi-layer networks. The number of recurrent and hidden units, and sequence length used varied for their tests. During subjective listening tests, the audio quality of the network was not deemed satisfactory by semiprofessional guitarists. Wright et al. test a new LSTM architecture along with another RNN variant, the Gated Recurrent Unit (GRU) in [33] and both are compared to the WaveNet architecture from [4]. The GRU, like LSTM, aims to remedy gradient problems by controlling information flow through the recurrent cell. The computations carried out in a GRU are similar to those of LSTM but with the various gates of the latter replaced with a single update gate for reduced complexity. The architecture used is comprised of a recurrent layer followed by a fully connected (FC) one, preliminary experiments showed that adding recurrent layers had little effect on the audio quality. This method was used to model a pedal (Big Muff) and a combo amplifier (Blackstar HT-1) [33]. In terms of objective quality, the most accurate RNN outperformed the WaveNet for the pedal and the most accurate WaveNet outperformed the RNN for the amplifier with LSTM outperforming the GRU in terms of accuracy with roughly the same processing time. In a follow-up article [32], further comparisons between the RNN from [33] and various WaveNet configurations were carried out for the modeling of two vacuum tube amplifiers: a Blackstar HT-5 Metal and a Mesa Boogie 5:50 Plus. These configurations were trained and compared in order to gauge the RT capabilities (in C++) and the audio quality. The LSTM provides better processing speeds than the WaveNet models however the largest WaveNet was better able to model the highly nonlinear HT5M amplifier.

Distortion effects can be emulated by various ANN architectures [28]. In this work, eight architectures are tested and compared including four LSTM networks, one hybrid convolutional LSTM, one MLP, one CNN, and one hybrid convolutional RNN. The most notable architectures of this work are: The parametric LSTM where the input dimensions are extended to take into account the amplifier parameters; The convolutional LSTM in which two 2D convolution layers are used to reshape the inputs for GPU parallelization; The sequence-to-sequence LSTM which outputs a buffer instead of a single sample. All architectures vary in terms of accuracy and RT performance but the hybrid convolutional LSTM presented the best Computation Time (CT) and accuracy trade-off. Another category of hybrid method includes autoencoders such as those presented in the works of Martinez-Ramirez et al. [26]. The general structure of an autoencoder comprises: an encoding front-end, a latent space, and a decoding back-end. This structure enables the model to learn an approximate copy of the input, forcing it to prioritize useful properties of the data. A convolutional autoencoder with FC latent space (dubbed CAFx) was presented with the following structure: A convolutional adaptive front-end, a latent-space of two dense layers, and a FC decoder. This architecture was used to model three effects of the IDMT-SMT-Audio-Effects dataset [17]: distortion, overdrive, and equalization (EQ). This architecture was modified by replacing the FC layers with Bidirectional LSTM (i.e. LSTM containing forward and backward information at every time step) to create a convolutional and recurrent autoencoder (CRAFx) to model more complex audio effects also from [17]. Another variant of this architecture uses a feedforward WaveNet in the latent space (CWAFx) and the three autoencoders are compared with the original WaveNet architecture [4] on various tasks including modeling a vacuum tube amplifier, sampled from a 6176 Vintage Channel Strip unit. The results showed that both the WaveNet and CAFx performed similarly but they are both outperformed by CRAFx and CWAFx with CRAFx performing slightly better than CWAFx. It was reported that the preamplifier was able to be successfully modeled on the two-second samples.

3.2   White-Box and Gray-Box Approaches

Neural networks are black-box by nature but in recent years they have started to be integrated into gray and white-box methods. The following works fall into the category of gray-box approaches. Parker et al. present the State Trajectory Network (STN) [24] which is a method of integrating neural networks into a State-Space model by adding circuit component values (e.g. voltages) to the training data for a more accurate simulation. A number of distortion circuits are tested, including a second-order diode clipper. Results show that this method is viable as all the circuits modeled were said to be indistinguishable from the targets in informal listening tests however the training can be unstable [24,25]. Aleksi Peussa augments the STN in his Masters thesis [25] to include recurrence using a GRU and compares this to both the original STN and a black-box GRU. This work confirms the STN's training instability as it was unable to model a Boss SD-1 pedal. Although the State-Space GRU was able to emulate this pedal, it was outperformed by its black-box equivalent. However the state-space model managed to outperform the black-box one when applied to a Moog ladder filter due to its self-oscillatory nature. Kuznetsov et al. [21] explore the idea of differentiable Infinite Impulse Response (IIR) filters using the Differentiable Digital Signal Processing (DDSP) library, which enables the integration of classic DSP elements in a differentiable setting [7]. The authors present the link between IIR filters and RNN and present a Wiener-Hammerstein model using differentiable IIR filters to emulate a Boss DS-1 distortion pedal, compared with a simple convolutional layer as a baseline. None of the models were able to fit the data perfectly using this method. Differentiable IIR filters are explored further as a cascade of differentiable biquads to model a distortion effect [22]. The proposed model is said to have significantly fewer parameters and reduced complexity compared to black-box ANN. This method was used to model a Boss MT-2 distortion pedal and comparison with WaveNet showed that the parametric EQ cascaded biquads outperformed the other representations as well as the WaveNet. Finally, white-box methods have also started to appear. Esqueda et al. [8] implement a white-box model in a differentiable form which allows approximate component values to be learned, thus remedying the accuracy problems of white-box modeling due to lack of access of exact component values. This method was tested on a Fender, Marshall, Vox tone stack as well as an Ibanez TS-808 Overdrive stage in order to validate the proposed model. The performance of any network no matter the category of approach is directly determined by a number of choices made regarding the training process. Notably, the choice of dataset is critical.

3.3   Datasets

The performance of ANN for any given task depends heavily on the dataset they are trained on. The network needs to be exposed to a wide range input-output pairings to generalize well. A number of datasets used for the task of amplifier modeling bear certain similarities. Almost all of the data used comprises clean guitar Direct Input (DI) sent through either the analog device or a

SPICE simulation. However, the data used depends on the approach. In black-box approaches only input-output recordings are necessary whereas for gray or white-box different or additional data is required. In the gray-box approaches presented [21,22,24,25], component values of the internal circuit are also used in the training data. In the white-box approach [8], only the circuit component values are used. In the black and gray-box approaches, certain aspects of the training data, namely the sampling rate, the length, and type of the data, have a significant impact on the resulting model. The sampling rate dictates the audio quality of the simulation and impacts its RT capabilities. The data used in [3] was obtained from a SPICE simulation of a Fender Bassman 56F-A preamplifier applied to DI from the Freesound dataset [9] at 44.1 kHz. A total of 4 h of data was used for the train set and 20 min for validation split into 100 ms segments and a random gain value was applied for more dynamic range. Damskägg et al. show that as little as three minutes of data is sufficient for the training of the CNN [4], although final results presented were obtained with five minutes of data (50% guitar and 50% bass) from the IDMT-SMT-Guitar/Bass datasets [18,19]. These datasets contain a variety of single note recordings of various different playing styles with varying pickups. The raw inputs were sent through three effects pedals: Ibanez Tube Screamer, Boss DS-1 and an Electro-Harmonix Big Muff Pi. This dataset was also used to train the RNN from [33]. The amplifier models of [32] used a different training set than the pedal emulation, taken from a pre-existing dataset. This dataset was tailor-made for this modeling task [29]. It includes five different styles of guitar sounds sent through various amplifiers with their gain parameters set to different levels. The audio used in [32] consists of around three minutes of guitar audio at 44.1 kHz with the training set consisting of 2 min 43 s. This data was used for both the WaveNet and the LSTM. The SignalTrain dataset [13] was used for train, test and validation of the shallower TCN architectures. This dataset contains input-output recordings at 44.1 kHz of various instruments from a LA-2A dynamic range compressor. The training data used for the NARX network of [2] was comprised of both signals from a function generator and an electric amplifier. All training data was recorded at 96 kHz. The recordings from this dataset were also used for the LSTM in [36]. As the choice of training data has a decisive impact on the performance of a neural network, so does the choice of cost function used in the optimization process.

3.4 Loss Functions and Evaluation Metrics

Training ANN is an optimization problem in which we often aim to minimize a given loss function representing the error between the prediction and the target. It must therefore accurately depict the perceptual difference between signals. This is often not the case with objective losses such as the Mean-Squared Error (MSE). MSE and similar time domain losses are computed directly on the signal's waveform which does not perfectly correspond to human perception. To improve the accuracy, spectral information can also be included but this approach presents its own problems and most of the losses used remain time domain-based. The most widely known, the MSE, is one of the most used for training ANN in distortion circuit simulation. It was used in one of the first articles presenting RNN for amplifier modeling [36] as well as in all of the gray-box methods presented previously, including a normalized variant used in [24] in order to stabilize the initial training of the network. Similar losses include the Root MSE (RMSE) used in [2] and the Normalized Root MSE (NRMSE) used in [28]. Despite having relatively widespread use, the MSE based losses lack perceptual accuracy. Another loss function frequently encountered is the Error-to-Signal Ratio (ESR) defined as:

$$\mathcal{L}_{\text{ESR}} = \frac{\sum_{i=0}^{N-1} |y[i] - \hat{y}[i]|}{\sum_{i=0}^{N-1} (y[i])^2} \tag{1}$$

Variants of this loss have been used in the works of Damskägg et al. [3,4] and Wright et al. [32,33] with pre-emphasis filtering for better perceptual accuracy. In Damskägg et al., the high-pass pre-emphasis filter with the transfer function $H(z) = 1 - -0.95z^{-1}$ was used to train their WaveNet as it was found that the model struggled at higher frequencies initially. To train the RNN, Wright et al. apply a different pre-emphasis filter with transfer function $H(z) = 1 - 0.85z^{-1}$ to the ESR along with a term to compensate for a DC offset in the prediction:

$$\mathcal{L}_{\text{DC}} = \frac{(\frac{1}{N} \sum_{i=0}^{N-1} (y[i] - \hat{y}[i]))^2}{\frac{1}{N} \sum_{i=0}^{N-1} (y[i])^2} \tag{2}$$

In a further work [34], various pre-emphasis filters and weightings are studied and compared to gauge which combination better reflects perceptual quality. The filters with the following transfer functions were tested: First-Order High-Pass ($H_{\text{HP}}(z) = 1 - 0.85z^{-1}$), Folded Differentiator ($H_{\text{FD}}(z) = 1 - 0.85z^{-2}$) and First-Order Low-Pass ($H_{\text{LP}}(z) = 1 + 0.85z^{-1}$). The low-pass filter is preceded by A-weighting which aims to mimic the equal loudness curves of the human ear. Listening tests showed that pre-emphasis filtering enabled better accuracy during modeling, with the A-weighted low-pass filtering achieving the best performance. The Mean Absolute Error (MAE) is also used in both a time domain formulation and a spectral variant. The MAE is defined as

$$\mathcal{L}_{\text{MAE}} = \frac{1}{N} \sum_{i=0}^{N-1} |\hat{y}[i] - y[i]| \tag{3}$$

and was used to train the autoencoders in [26]. Steinmetz et al. [31] use a combination loss comprising both time domain, using MAE, and spectral features. For the spectral magnitude, the Short-Term Fourier Transform (STFT) loss [1] is used, leading to the following cost function ($\|.\|_{\text{F}}$ the Frobenius norm): $\mathcal{L}_{\text{MAE}} + \mathcal{L}_{\text{STFT}}$ with $\mathcal{L}_{\text{STFT}} = \mathcal{L}_{\text{SC}} + \mathcal{L}_{\text{SM}}$ where

$$\mathcal{L}_{\text{SC}} = \frac{\||\text{STFT}(y)| - |\text{STFT}(\hat{y})|\|_{\text{F}}}{\||\text{STFT}(y)|\|_{\text{F}}} \text{ and} \tag{4}$$

$$\mathcal{L}_{\text{SM}} = \frac{1}{N} \|\log(|\text{STFT}(y)|) - \log(|\text{STFT}(\hat{y})|)\|_1, \tag{5}$$

The loss functions used for training can also be applied to the evaluation process for an objective measure of performance. These losses must be differentiable to be used in the gradient-based optimization. Non-differentiable functions and subjective listening tests constitute other methods of evaluation to provide a more comprehensive assessment of the quality during testing. While objective metrics struggle to properly reflect the perceptual aspects of the output, listening tests take time to implement and hinder continuous integration of ML systems. Therefore, there is need for objective evaluation metrics. Most of the evaluation methods used for this task rely on reuse of the loss functions used during training or a variation thereof. Damskägg et al. [4] use pre-emphasized ESR for training and plain ESR for evaluation. While simple to implement, a single objective metric cannot replace subjective listening tests. The listening tests that are mainly used for this task rely on MUltiple Stimuli with Hidden Reference and Anchor (MUSHRA) testing. In MUSHRA, the participants are presented with a labeled reference, various test samples, an unlabeled reference, and an anchor [14]. A similar framework used to carried out listening tests with human participants is the Web Audio Evaluation Tool [16] based on the HTML5 Web Audio API for perceptual audio evaluation. For the objective evaluation of the models compared in [31], three metrics were used: the MAE and the STFT loss described above and a perceptually informed loudness metric that uses the loudness algorithm from the ITU-R BS. 1770 Recommendation [15]. A listening test similar to MUSHRA was also carried out using the WebMUSHRA interface [30] to further validate the accuracy of the models. The results of this test indicated that a small difference was perceived among the models in comparison to the reference. WebMUSHRA was again used in [32] and showed the largest WaveNet to be the most perceptually accurate model of the HT5M amplifier, although this prediction could still be distinguished from the original amplifier. The largest LSTM proved to be the closest to the Mesa 5:50 amplifier in terms of subjective quality and most people could not tell the difference between the model and the target. For the work established in Thomas Schmitz's PhD thesis [28], a number of objective metrics as well as listening tests were presented. One listening test studies the number of parameters that can be reduced without loss of accuracy and the other is used to determine the threshold of a given metric above which the accuracy is no longer improved. An overview of the evaluation methods is presented in here: **Objective:** MSE [36], RMSE [2,28], NRMSE [28], Spectrogram [28], Power Spectrum [28], Harmonic Analysis using ESS [28], Waveform plot [28] [32,33], SNR [28], Difference in harmonic content [28], MAE [26], ESR [3,4,32,33], Frequency spectrum [3,4], MS_MSE [26], MFCC_COSINE [26], Spectra of 1245-Hz sinusoid to study aliasing [4], AME [25], CT [3,4,28,32], Custom metric taking into account RMSE + CT [28]. **Subjective:** Aural comparison of prediction and target [36], 2 listening tests [28], MUSHRA [3,32], Web audio evaluation tool [26], pre-emphasized ESR [34].

The audio quality of the prediction is not the only aspect that requires evaluation. The RT capabilities are crucial to take into account when studying VA models and can also be presented as an objective metric of the emulation quality.

This is illustrated in the last two objective methods listed above that take into account the CT.

3.5 Real-Time Capabilities

A major factor to take into account in VA modeling is the RT constraint. If the simulation cannot process in RT then it is of little use to the users. The RT constraint for this application is approximately 10 ms. Any latency above this is likely to be perceived by the musician and hinder their playing. The latency produced in digital audio is not limited to only the CT of the Digital Signal Processing (DSP) algorithm used which significantly decreases the time left for the DSP computations. This means that using DL architectures with a lot of computations is complicated as it either entails too high a latency or excessive CPU usage. A number of the architectures are capable of RT to a degree but caveats also exist. The architectures although often capable of RT processing, some of which are light-weight enough to work in RT on CPU, are still computationally heavy and have not been demonstrated to be able to achieve RT speeds for sampling rates over 44.1 kHz. The hybrid convolutional and recurrent architecture from [28] utilizes GPU parallelization to process the data in RT which poses a problem when CPU processing is required. The autoencoders [26] have been demonstrated using a sampling rate of 16 kHz which is insufficient for high-quality audio applications. The shallow TCN [31] are capable of RT use only for buffer sizes over 1024 samples, which at 44.1 kHz incurs a latency of around 23.2 ms, over double the RT constraint of 10 ms. Finally the architectures of [32], although capable of RT processing, remain computationally heavy, even at 44.1 kHz which is the lower bound for high quality in music applications. Digital implementations can introduce aliasing. To remedy this, anti-aliasing techniques are used which often require upsampling the signals by a factor of eight [35] which greatly increases the number of samples to be processed, further restraining the allowed CT of the DSP algorithm. This also applies to ANN although formal study of the aliasing introduced is lacking.

Here we present an overview of the RT capabilities of the architectures in the state-of-the-art in terms of their Real Time Factor (RTF) with $RTF = \frac{Processing\ Time}{RT\ constraint}$. RTF lower than 1 is required for RT operation. The sampling rate used is 44.1 kHz unless stated otherwise: The two-layer LSTM and the three-layer MLP [28] have a RTF of 1.39 and 0.24 respectively on GPU. The WaveNet amps [3] had RTF equal to 0.16 (WaveNet1) & 0.33 (WaveNet3) in Python whereas the WaveNet pedals [4] had RTF of 0.53 (WaveNet1) & 0.91 (WaveNet3) in C++. The CRAFx architecture [26] had RTF of 1.44 in Python on CPU at 16 kHz. The single-layer LSTM & GRU [33] had RTF of 0.097 for the fastest RNN & 0.41 for the slowest, estimated on CPU. And the CWAFx [26] had 1.48 in Python on CPU at 16 kHz. Finally, the shallow TCN [31] was capable of RT for large frame sizes in Python on CPU. A number of architectures are capable of RT use, even on CPU. However, the RT measures presented in here vary in a number of ways including: the sample rate used, the processing unit, the implementation language, and the RTF definition (number of operations, timing the inference, etc.). This makes formal comparison challenging.

4 Discussion

Overall, the wide range of parameters used in the state-of-the-art make a formal comparison of the architectures complicated. To remedy this, we chose at least one model from each class of networks and trained them using the same data and loss function. This allows for a clearer comparison. The dataset used for training is from [29], using 80% for the training data and 20% for testing. The sampling rate used is 44.1 kHz and the comparison is limited to black-box approaches. The loss function is ESR without pre-emphasis filtering. A DC term was added to the ESR when training the recurrent network as the predictions from model are known to have an amplitude offset. The training parameters for each network are the following: LSTM: Sequence-to-sequence LSTM with 32 recurrent units; WaveNet: Number of channels = 12; dilation depth = 10; kernel size 3; Convolution LSTM: Number of channels = 35; stride = 4; kernel size = 3 (for both convolution layers); Shallow-TCN: Number of channels = 32; number of blocks = 4; stack size = 10; dilation growth = 10. The Shallow TCN was trained for 150 epochs instead of the original 60 to account for the difference in training data and loss and all the other networks were trained until an early stopping condition was met. For each model tested, we present a number of objective metrics as well as the inference speeds. The STFT reported is the Aggregate STFT [1]. Table 1 shows that the Shallow TCN outperforms the other architectures in terms of processing speed but for significantly lower objective quality. The lower quality could be due to insufficient training as this network was trained for a fixed number of epochs. The LSTM with 32 units outperforms all other models in terms of objective measures but is outperformed by the CNN in terms of processing speed. This is contradictory to the results presented in Wright et al. [32] which show that most LSTM models were able to outperform the WaveNet models in terms of CT. This difference could be due to the evaluation method as Wright et al. report CT of an optimized C++ implementation of both networks and the results presented here were all obtain using Python. The hybrid conv-LSTM network from Schmitz et al. [28] ranks highly in terms of

Table 1. Black-box architecture comparison. We define the Real-Time factor here to be RTF = $\frac{\text{Processing Time}}{\text{RT constraint}}$. RTF lower than 1 is required for RT operation. All results reported were recorded in Python on CPU (AMD Ryzen 7 3750H CPU at 2.3 GHz).

Architecture	RTF	MSE	ESR	MAE	STFT
RNN (LSTM-32) [33]	0.51	0.0040	0.0244	0.0378	0.5952
CNN (WaveNet) [3]	0.35	0.0703	0.4337	0.1359	0.6542
Hybrid (Conv-LSTM) [28]	3.25	0.0069	0.0423	0.0530	0.6937
CNN (Shallow TCN) [31]	0.14	0.3190	2.1371	0.4510	1.2348

objective quality but is incapable of RT use without GPU parallelization. Over-all, all models except for the Shallow-TCN were reported to have acceptable subjective quality, judged through informal listening tests, even though their objective time domain metrics vary. This highlights the divergence between the objective metrics and perceptual quality. The STFT loss produces values closer to perception, showing that taking into account spectral features can improve the perceptual accuracy. All the audio results are available at https://www.math.u-bordeaux.fr/~plegra100p/NNA_AMPLI_EMU.php.

The architectures presented here overall are capable of accurately modeling analog distortion effects. However, some aspects warrant further study. These architectures are often capable of RT processing, some of which are light-weight enough to work in RT on CPU. However, they remain computationally heavy and have not been demonstrated to be able to achieve RT speeds for sampling rates over 44.1 kHz. Indeed, some models either require parallelization of their operations on GPU, low sampling rates or large buffer sizes in order to achieve close to RT performance. Furthermore, the architectures currently present in the state-of-the-art that are capable of RT use have only be demonstrated to work with sampling rates of 44.1 kHz which is the lower bound for high quality audio in music applications. Moreover, digital implementations of analog audio effects usually introduce aliasing into the signal and to remedy this, anti-aliasing techniques are used which often require upsampling of the signals by a factor of 8 [35]. In [4], Damskägg et al. study the effect of aliasing in the prediction of their WaveNet and claimed that aliasing was indeed present in the output even though the models were trained on non-aliased data but that this aliasing could not clearly be heard in the predictions. Therefore anti-aliasing techniques might be required for neural models and further work on the possible impact of aliasing should be explored. A variety of methods presented here allow for input parame-ters to be taken into account in the network for a parametric model via an extra input dimension. However these methods could slow down training significantly and greatly increase the amount of data needed. Moreover, all mentions of these parametric approaches in the literature have been hypothetical or implemented with marginal success and no clear demonstration of the methods have been presented that we know of. The cost functions used for training and evaluation of the networks have been studied in recent years. Wright et al. [34] present a study on various functions for pre-emphasis filtering and weighting of the signal in order to better capture the perceptual features. It was shown that the loss that best improved audio results was the ESR with low-pass and A-weighting pre-emphasis. A more formal comparison of various cost functions for amplifier modeling would be desirable.

5   Conclusion

In this work, we present an overview of the current state-of-the-art of neural network-based VA modeling, covering the recent advances in deep learning in this field under black-box, gray-box and white-box approaches. We highlight the results of each method, including the audio quality and RT capabilities. Moreover, we include the evaluation methods used and the limitations of each method. This was done in order to identify possible avenues for further work. We showed that RT capabilities, and possible aliasing of such approaches as well perceptually relevant and preferably differentiable objective metrics warrant further investigation. Further exploration into parametric models to enable adjustment of the amplifier settings is also desirable.

References

1. Arik, S.Ö., Jun, H., Diamos, G.: Fast spectrogram inversion using multi-head convolutional neural networks. Inst. Electr. Electron. Eng. (IEEE) **26**, 1–6 (2018)
2. Covert, J., Livingston, D.L.: A vacuum-tube guitar amplifier model using a recurrent neural network. In: Proceedings - IEEE SouthEactCon December, 1–5 (2013)
3. Damskägg, E.P., Juvela, L., Thuillier, E., Välimäki, V.: Deep learning for tube amplifier emulation. In: Proceedings of the IEEE International Conference on Acoustics Speech Signal Process. (ICASSP-19), Brighton, UK, May 2019, pp. 471–475 (2019)
4. Damskägg, E.P., Juvela, L., Välimäki, V.: Real-time modeling of audio distortion circuits with deep learning. In: Proceedings of the SMC Conferences, pp. 332–339 (2019)
5. Eichas, F., Möller, S., Zölzer, U.: Block-oriented gray box modeling of guitar amplifiers. In: Proceedings of the 20th International Conference on Digital Audio Effects, pp. 184–191 (2017)
6. Eichas, F., Zölzer, U.: Black-box modeling of distortion circuits with block-oriented models. In: Proceedings of the 19th International Conference on Digital Audio Effects, DAFx September, pp. 39–45 (2016)
7. Engel, J., Hantrakul, L., Gu, C., Roberts, A., Team, B., View, M.: DDSP: differentiable digital signal processing. Digital Signal Process. 1–19 (2020)
8. Esqueda, F., Kuznetsov, B., Parker, J.D.: Differentiable white-box virtual analog modeling. 24th International Conference on Digital Audio Effects, pp. 41–48 (2021)
9. Fonseca, E., et al.: Freesound datasets: a platform for the creation of open audio datasets. In: International Society for Music Informatic Retrieval (ISMIR) (2017)
10. Fractal Audio Systems: Multipoint Iterative Matching and Impedance Correction Technology (MIMIC ^{TM}). Technical report, April, Fractal Audio Systems (2013)
11. Gers, F.A., Schmidhuber, J., Cummins, F.: Learning to forget: continual prediction with LSTM. Neural Comput. **12**(10), 2451–2471 (1999)
12. Gillespie, D.J., Ellis, D.P.: Modeling nonlinear circuits with linearized dynamical models via kernel regression. In: IEEE Workshop on Applications of Signal Processing to Audio and Acoustics, vol. 8 (2013)
13. Hawley, S., Colburn, B., Mimilakis, S.I.: SignalTrain LA2A dataset. In: Audio Engineering Society 147th Conference (AES 147) (2020)

14. ITU-R: ITU-R BS.1534-3: Method for the Subjective Assessment of Intermediate Quality Level of Audio Systems. International Telecommunication Union Radiocommunication. Sector BS Series, vol. 34 (2015)
15. ITU-R BS.1770-4: Algorithms to Measure Audio Programme Loudness and True-peak Audio Level BS Series Broadcasting Service (sound). International Telecommunication Union Radiocommunication Sector 4 (2015)
16. Jillings, N., Moffat, D., De Man, B., Reiss, J.: Web audio evaluation tool: a browser-based listening test environment. In: SMC (2015)
17. Kehling, C., Männchen, A., Eppler, A.: IDMT-SMT-Audio-Effects. Technical report, Fraunhofer Inst. for Digital Media Technology IDMT (2010)
18. Kehling, C., Männchen, A., Eppler, A.: IDMT-SMT-Guitar. Technical report, Fraunhofer Inst. for Digital Media Technology IDMT (2014)
19. Kehling, C., Männchen, Andreas, Eppler, A.: IDMT-SMT-Bass. Technical report, Fraunhofer Inst. for Digital Media Technology IDMT (2010)
20. Kemp, M.J.: Analysis and Simulation of Non-Linear Audio Processes using Finite Impulse Responses Derived at Multiple Impulse Amplitudes. The 106th AES Convention p. Preprint no.4919 (1999)
21. Kuznetsov, B., Parker, J.D., Esqueda, F.: Differentiable IIR filters for machine learning applications. In: 23rd International Conference on Digital Audio Effects, pp. 297–303 (2020)
22. Nercessian, S., Sarroff, A., Werner, K.J.: Lightweight and interpretable neural modeling of an audio distortion effect using hyperconditioned differentiable biquads. In: Proceedings of the ICASSP, IEEE International Conference on Acoustics, Speech and Signal Processing 2021-June(2), pp. 890–894 (2021)
23. van den Oord, A., et al.: WaveNet: a generative model for raw audio. arXiv preprint arXiv:1609.03499 1–15 (2016)
24. Parker, J.D., Esqueda, F., Bergner, A.: Modelling of nonlinear state-space systems using a deep neural network. In: Proceedings of the 22nd International Conference on Digital Audio Effects, DAFx, pp. 165–172 (2019)
25. Peussa, A.: State-Space Virtual Analog Modelling of Audio Circuits. Masters thesis, Aalto University, p. 73 (2020)
26. Ramírez, M.A.M., Benetos, E., Reiss, J.D.: Deep learning for black-box modeling of audio effects. Appl. Sci. (Switzerland) 10(2), 638 (2020)
27. Rugh, W.J.: Nonlinear System Theory: The Volterra/Wiener Approach. The Johns Hopkins University Press, Baltimore (1981)
28. Schmitz, T.: Nonlinear Modeling of the Guitar Signal Chain Enabling its Real-time Emulation. Ph.D. thesis, University of Liège October, 258 (2019)
29. Schmitz, T., Embrechts, J.J.: Introducing a dataset of guitar amplifier sounds for nonlinear emulation benchmarking. In: AES E-Library, pp. 1–4 (2018)
30. Schoeffler, M., Bartoschek, S., Stöter, F.R., Roess, M., Edler, B., Herre, J.: web-MUSHRA - a comprehensive framework for web-based listening tests. J. Open Res. Software 6, 8 (2018)
31. Steinmetz, C.J., Reiss, J.D.: Efficient neural networks for real-time analog audio effect modeling. arXiv preprint arXiv:2102.06200 (2021)
32. Wright, A., Damskägg, E.P., Juvela, L., Välimäki, V.: Real-time guitar amplifier emulation with deep learning. Appl. Sci. (Switzerland) 10(3), 766 (2020)
33. Wright, A., Damskägg, E.P., Välimäki, V.: Real-time black-box modelling with recurrent neural networks. In: Proceedings of the 22nd International Conference on Digital Audio Effects (DAFx-19) September, 1–8 (2019)

34. Wright, A., Välimäki, V.: Perceptual loss function for neural modeling of audio systems. In: Proceedings of the ICASSP, IEEE International Conference on Acoustics, Speech and Signal Processing 2020-May, pp. 251–255 (2020)
35. Yeh, D.T.: Digital Implementation of Musical Distortion Circuits by Analysis and Simulation. Ph.D. thesis, Stanford University (2009)
36. Zhang, Z., Olbrych, E., Bruchalski, J., McCormick, T.J., Livingston, D.L.: A vacuum-tube guitar amplifier model using long/short-term memory networks. In: Conference Proceedings - IEEE SOUTHEASTCON 2018-April, 1–5 (2018)

Defining a Quality Measure Within Crossover: An Electric Bus Scheduling Case Study

Darren M. Chitty[✉] and Ed Keedwell

College of Engineering, Mathematics and Physical Sciences, University of Exeter,
Exeter EX4 4QF, UK
darrenchitty@gmail.com, E.C.Keedwell@exeter.ac.uk

Abstract. Genetic Algorithm (GA) crossover for permutation type problems is difficult due to the avoidance of vertex or value repetition. As a result extensive research into crossover operators has been undertaken with many variants developed. However, these crossover operators operate in a *blind* manner relying on the mechanics of survival of the fittest. A possible improvement is to introduce a *quality* measure into crossover enabling high quality edges to be utilised. This paper presents a crossover operator ER-Q that selects parental edges based upon their *quality* and applies this to an electric bus scheduling problem. Results demonstrate significant improvements in electric bus scheduling over alternative *blind* crossover operators. This paper also explores the definition of *quality* in terms of the electric bus scheduling problem noting that quality is difficult to quantify. A range of quality metrics are presented that can be used with differing effectiveness to optimally schedule electric buses.

Keywords: Genetic Algorithms · Electric Bus Scheduling · Crossover

1 Introduction

Meta-heuristic methods such as a Genetic Algorithm (GA) [14] are popular for application to permutation routing problems such as the Traveling Salesman Problem (TSP), Vehicle Routing Problem (VRP) or a bus scheduling problem. Permutation problems are $\mathcal{NP}$-hard in nature and complex as each vertex must occur once only. GAs use the principles of Darwinian evolution, survival of the fittest, mutation and crossover to improve a population of solutions. Crossover, whereby genetic material is swapped between parent solutions, is problematic when avoiding vertex repetition. Consequently, many differing crossover operators have been developed to address this issue. Some preserve parental paths [9,12,20], some parental edges [28] and some use a graph-based approach [19,27]. However, these crossover operators have one common factor in that they rely on the mechanics of evolution to progressively improve solutions. They assume that parent solutions are highly fit and thus all edges must be too. Furthermore, to resolve vertex conflicts, new edges are introduced which will have an unknown effect on solution quality. In effect, these crossover operators operate in a *blind* manner relying on survival of the fittest to remove poorly recombined solutions.

This paper presents a methodology for less *blind* crossover whereby the operator is augmented with a *quality* measure to analyse edges. This enables better quality edges to be selected and more importantly, the avoidance of violating problem constraints. To test this hypothesis the proposed quality-based ER-Q crossover will be compared to a range of *blind* crossover operators using a complex electric bus scheduling problem which has the constraints of a timetable and bus range. Furthermore, this paper will consider the definition of quality itself which can be difficult to quantify. A wide range of quality metrics could be used and given the locality of its use, an edge by edge basis, choice of metric is shown to be important in terms of overall results. The paper is laid out as follows: Sect. 2 will provide an overview of crossover, its drawbacks and prior work that considers edge quality; Sect. 3 will introduce a quality based crossover technique and its application to an electric bus routing problem. In Sect. 4, the results of a range of crossover operators including the proposed quality-based operator when applied to a real-world electric bus scheduling problem will be presented. Definitions of quality for electric bus scheduling will be explored and tested in Sect. 5 and finally Sect. 6 will sum up and draw conclusions.

2   Background and Related Work

Permutation type routing problems have solutions which contain non-repeated values. Examples include the Traveling Salesman Problem (TSP) with the aim to visit each city once only in the shortest distance or the Multi Depot Vehicle Routing Problem (MDVRP) [8] with the aim to assign customers and routes to a fleet of vehicles minimising distance. The Capacitated Vehicle Routing Problem (CVRP) extends the MDVRP adding a vehicle capacity such as a weight limit. The bus scheduling problem can be considered similar to the MDVRP with time windows (MDVPTW) whereby the goal is to assign a set of timetabled routes to buses such that the bus fleet traversal distance is minimised whilst performing each route on time. The electric bus scheduling problem adds a further constraint in that each bus has a limited range due to its battery which cannot be quickly recharged. Therefore, an electric bus must be able to perform its assigned timetabled routes and return to the depot without running out of charge.

Meta-heuristic approaches are commonly used to solve permutation type problems such as bus scheduling. A popular meta-heuristic, the Genetic Algorithm (GA) [14], uses the principles of Darwinian evolution to derive solutions maintaining a population of solutions generated using natural selection, crossover and mutation. Parents are probabilistically selected based on their solution quality and their genetic encoding used to generate offspring. This is achieved using crossover whereby genetic material between parents is swapped to create two new solutions. Given the success of GAs they have been applied to bus scheduling problems. For instance, Kidwai et al. [17] use a GA to minimise the fleet size of buses for a set of timetabled routes in Burdwan, India. For electric bus routing Janovec and Kohani [16] used a grouping GA, whereby bus route assignments are termed groups, to minimise the energy use of a bus fleet. Wang et al. [25] consider a multi-depot three line electric bus routing problem from Qingdao China using a column generation GA whereby columns representing allocated routes for electric buses are recombined. Hu et al. [15] optimally route electric buses with additional fast bus stop charging using a Mixed Integer Linear Programming (MILP) model and a GA for a three bus route in Sydney.

However, crossover of genetic material between parents can be problematic for permutation problems such as bus scheduling. Vertices must occur once only but if a vertex occurs at differing points in parent chromosomes it could occur twice in a child. Hence, specialised GA crossover operators have developed for solving permutation problems early attempts preserving parent sub-paths. *Order* (OX) [9] takes a parent subsequence and preserves the relative order of vertices from the second copied to the child in order from the second crossover point skipping conflicts. *Partially Mapped* (PMX) [12] transfers vertices between crossover points directly to children. Each child takes vertices from the other parent outside these points resolving conflicts using a one to one mapping of crossed over genetic material. *Cyclic* (CX) [20] ensures that each vertex and relative position comes from a parent using a one to one mapping. The vertex from position one in parent one is directly copied to the child. The next is the corresponding vertex in parent two but in the position it is found in parent one. This continues until a conflict when genetic material from the second parent can be directly copied.

More recent research on crossover considers edges between vertices more important. Grefenstette [13] introduced a method whereby from a given vertex only incident edges from either parent could be selected probabilistically based on length. If none are available a random edge is selected. Edge Recombination (ER) [28] extended the concept by constructing an edge list from two parents with each vertex having between two and four edges. A solution is constructed by taking an initial vertex and selecting edges from the edge list which have least onward edges. If no parent edges are available a random edge is used.

The aforementioned operators assume parental paths or edges are better quality due to natural selection. However, parents will likely have some lower quality edges thus these operators used in a purely evolutionary way fail to reach optimality. To improve solutions local search methods are often used. For example 2-opt [7] which iteratively takes all solution edge pairs and swaps them, a sub-tour reversal accepting if an improvement. Given local search success and importance of edge retention used by ER, a new more successful direction of crossover developed with Edge Assembly (EAX) [19]. This operator takes two parents A and B and constructs a graph G that contains the edges of parents. From G a set of AB-cycles are constructed, an even-length sub-cycle of G with edges alternating from A and B whereby cities can repeat but not edges. A subset of AB-cycles are selected as an E-set. Once these have been determined an intermediate solution is created from this E-set by using a greedy local search.

Edge preservation with local search has become the dominant crossover applied to the TSP. Partition crossover (PX) [26] uses the theory that if parent solutions are locally optimal, offspring from their preserved edges are likely locally optimal too *tunnelling* to new optima. PX creates a graph G of parental edges finding a partition that separates graph edges aside from two edges. Off-spring are generated using one parental sub-tour from one partition and a sub-tour from the other partition and parent. PX cannot introduce new edges so requires additional methods but achieved good results. PX was extended by Generalised Partition crossover (GPX) [27] which utilises all partitions with only two connecting edges to construct offspring. Combined with Lin-Kernighan local search performance was similar to Chained Lin-Kernighan [2]. GPX was improved with further partition recombining methods (GPX2) [24].

However, a common factor of these crossover operators is that they operate *blindly* without problem domain knowledge or edge *quality* awareness. It could be argued that relying on the dynamic of pure Darwinian evolution by using a *blind* crossover is not particularly beneficial. Indeed, Osaba et al. [21] describe the CX crossover as *blind* and through analysis of a TSP problem postulate that CX provides little benefit, mutation and natural selection are the major contributing factors. The heuristic of Grefenstette [13] considered parent edge *quality* using distance but reported edge failures of 40%. However, edge quality can be implicitly defined, Tang and Leung [22] considered using nearest neighbours to select edges in cases of failure of parent edge availability. Edges between nearest neighbours are naturally of higher quality. Alternatively, Ting [23] modifies ER crossover such that edges from alternating parents are used with edge failures resolved by greedily selecting the shortest available edge. Freisleben and Merz [11] use a greedy approach to solve TSPs whereby non-common parental edges are deleted in offspring and reconnections made using the shortest available edge. Kkesy and Domański [18] considered an alternative edge *quality* measure, the *sensitivity* of an edge in a parent to being broken and essentially replaced by two edges with an intermediate vertex. Parental edges of higher sensitivity are more likely to be preserved. Ahmed [1] modified ER crossover to consider edge quality by greedily using the parental edge from a vertex with the best quality. Results from application to small TSP instances yielded improvements over ER.

An alternative meta-heuristic Ant Colony Optimisation (ACO) [10] integrates problem domain knowledge to solve routing problems. ACO uses *heuristic* information, or edge *quality*, alongside pheromone information to guide ants constructing solutions by traversing graph G. For the TSP this quality is the distance between vertices. This concept of heuristic information can be successfully incorporated into GA crossover via ACO. Branke et al. [3] integrated ACO into a crossover operator (ABX). A temporary pheromone matrix is created from a number of parents and combined with heuristic information to generate a set of offspring solutions. An alternative approach combined PMX crossover with ACO (ACOX) whereby ants resolve conflicts rather than using a one to one mapping hence incorporating edge quality improving upon PMX considerably [4].

3   Embedding a Quality Measure Within Crossover

To apply a meta-heuristic GA to the electric bus scheduling problem a chromosome will represent a solution with a set of buses each followed by a set of timetabled routes to be completed in order. However, as discussed in Sect. 2, *blind* crossover methods such as OX, ER and PMX, with no knowledge of the bus scheduling problem and its constraints could be deemed to be less effective. A methodology is required that favours preservation of parental edges but selects edges using *quality* and non-violation of constraints. Consequently, a novel crossover operator based upon ER but incorporating an edge quality measure, ER-Q [6], is proposed for application to the electric bus scheduling problem. ER crossover aims to preserve parental edges building solutions step by step using any available parental edges. From a given vertex ER selects the parental edge with fewest onward connections to avoid *edge failures* when no parental edges are available. In this instance a random choice is made over all available edges.

$$p_1 = (1\,2\,3\,4\,5\,6\,7\,8\,9\,0)$$
$$p_2 = (7\,4\,1\,6\,0\,5\,2\,8\,3\,9)$$

Given two parents p_1 and p_2 an initial vertex 1 is selected. The next vertex is taken from the connecting edges 0, 2, 4, or 6. All four have three available parental edges so 4 is randomly selected. From vertex 4 edges to vertices 3, 5 and 7 are available, all have three available parental edges so 7 is randomly selected. From vertex 7 only two parent edges are available to vertices 6 and 8 with 6 having fewest parental edges so is automatically selected giving the following:

$$o = (1\,4\,7\,6\,x\,x\,x\,x\,x\,x)$$

This continues until a solution comprised mainly of parental edges is derived. The ER crossover operator is described in Algorithm 1. *Edge failures* of 1–1.5% are reported by the authors, a weakness with *random* edge choices made introducing possibly poor edges which violate problem constraints. Even selecting a random parental edge could violate problem constraints or be a poor quality edge.

Algorithm 1. Edge Recombination Crossover

1: E = list of parent edges from each vertex, K = empty list, N = random vertex
2: **while** length of K is less than length of parent **do**
3:     append N to K, remove N from parent edge list E
4:     **if** parental edge list E at N not empty **then**
5:         N = vertex from E with fewest connections (random choice if multiple)
6:     **else**
7:         N = random available vertex (an edge failure)
8:     **end if**
9: **end while**

ER-Q can resolve the problem of violating problem constraints or inserting a poor quality edge by embedding a constraint and edge quality aware measure. Since ER builds a complete solution step by step constraints such as an electric bus charge level and current time can be tracked for each bus such that edges to routes that result in running out of charge or lateness can be considered taboo. Of the remaining edges, a high quality edge can be selected, a less *random* approach. The highest quality edge may not be the optimal so a probabilistic model over edges is used, similar to ACO when ants probabilistically decide vertices to visit using the *random proportional rule*, the probability of ant k at vertex i visiting vertex $j \in N^k$ is defined as:

$$p_{ij}^k = \frac{[\tau_{ij}]^\alpha [\eta_{ij}]^\beta}{\sum_{l \in N^k} [\tau_{il}]^\alpha [\eta_{il}]^\beta} \tag{1}$$

where $[\tau_{il}]$ is pheromone on vertex i to vertex l edge; $[\eta_{il}]$ is heuristic information, $1/d_{il}$; α, β are tuning parameters; N^k is the feasible neighbourhood of ant k.

When an edge failure occurs at vertex i, heuristic information can be similarly used for edge probability selection. A simple quality measure for the electric bus scheduling problem and minimising fleet distance would be to use the distance from the end of one bus route to the start of the next, vertex i to vertex j, d_{ij}. If assigning a bus a route represented by vertex j violates the energy use or lateness constraints, the edge has zero probability of selection. Thus, with ER-Q, the probability of taking the edge to vertex $j \in N$ can be defined as:

$$p_{ij} = \frac{[\eta_{ij}]^\beta [T_{ij}]}{\sum_{l \in N} [\eta_{il}]^\beta [T_{il}]} \tag{2}$$

where $[\eta_{il}]$ is heuristic information, $1/d_{il}$; β tunes edge length importance; N is the feasible neighbourhood; T_{il} indicates if the edge from vertex i to l is taboo:

$$T_{il} = \begin{cases} 0 & \text{if current time plus time to travel to vertex } l \text{ causes lateness} \\ & \text{or energy required to travel to vertex } l \text{ and back to the depot} \\ & \text{exceeds remaining battery charge} \\ 1 & \text{otherwise} \end{cases} \tag{3}$$

This hypothesis of selecting edges based on *quality* can be further extended to parental edge selection. Now edge quality is accounted for in cases of *edge failure* and thus less problematic, it can be hypothesised that the ER policy of selecting parent edges with fewest onward edges is unnecessary. If parental edges are available then a probabilistic decision of which to take can also be made based purely upon their edge *quality*. In effect, the parent solutions act as a *candidate* set of edges. The quality aware ER crossover operator, ER-Q, is described in Algorithm 2. Note the key differences to the standard ER crossover operator in Algorithm 1. Now, for *edge failures* a probabilistic model using edge quality is used to select an available edge detailed on lines 8–9 and to select a parental edge a probabilistic model using edge quality is constructed over just the parental edges, lines 5–6.

Further enhancements can be applied to ER-Q, minimising the degree of modification of parent solutions and forcing edge failures [6]. Given the nature of using a probabilistic model, making many decisions can inevitably lead to occasional errors in these decisions. To avoid this, each parent contributes directly to an offspring whereby a random set of bus routings are directly preserved. This is similar to standard crossover whereby the genetic material outside two points of a parent solution is preserved. The remaining bus routings are constructed as described in Algorithm 2. The second enhancement is to in effect cause an edge failure with

---

Algorithm 2. ER-Q Crossover

```
 1: E = list of parent edges from each vertex, K = empty list, N = random vertex
 2: while length of K is less than length of parent do
 3:     append N to K, remove N from parent edge list E
 4:     if parental edge list E at N not empty then
 5:         create probabilistic model over edges from E using quality measure
 6:         N = select vertex using random proportional rule (Equation 2))
 7:     else
 8:         create probabilistic model over all available edges using quality measure
 9:         N = select vertex using random proportional rule (Equation 2))
10:     end if
11: end while
```

---

a given low probability even if parental edges are available. Now problem constraints and edge quality are accounted for using the probabilistic model, edge failures can be considered less problematic. Indeed, consideration of non-parental edges can now introduce new high quality edges into the population.

4 Optimally Routing an Electric Bus Fleet

To test the theory that the less *blind* crossover operator ER-Q can improve evolution it will be applied to an electric bus scheduling problem. A UK bus operator runs buses throughout a large area consisting of a radius of 50 km. These run on a range of routes whereby a number of buses operate to a given timetable. The bus operator uses a fleet of electric buses which are equipped with a 450 kWh battery providing a range of 185 km using 2.42 kWh of energy per Km. The objective is to assign timetabled routes to electric buses such that the distance the fleet travels is minimised with no tardiness or violating the range constraints of the buses. A solution to the electric bus scheduling problem will consist of a set of unique values representing buses each followed by a set of unique values representing its assigned routes. An electric bus performs their assigned routes in the given order. A set of routing scenarios have been created from the UK bus operator routes varying in size and are described in Table 1.

Table 1. Real-world electric bus routing problems.

Problem	Lines	Routes	Buses	Distance (Km)	Problem	Lines	Routes	Buses	Distance (Km)
Scenario A	20	253	150	3112.58	Scenario F	24	670	150	7244.83
Scenario B	20	223	150	2353.76	Scenario G	60	1456	450	21050.11
Scenario C	20	890	150	13555.99	Scenario H	64	1774	450	29836.56
Scenario D	20	518	150	12145.38	Scenario I	124	3230	900	50886.67
Scenario E	20	676	150	12474.52					

To test the effectiveness of the quality-based crossover ER-Q comparisons will be made with OX, CX, PMX, ER and PX crossover operators. Three simple mutation operators, swap, insert and inversion will be used. Note that to be able to fully assess the effectiveness of each crossover operator no local search methods will be used. Experiments were conducted over 25 random runs using a Ryzen 2700 processor and a parallel method to maximise CPU occupancy [5]. The parameters used for the GA implementation are described in Table 2. A recommended low degree of permissible maximum bus route modification of 10 bus routes is used and a one in ten probability of forcing edge failures [6].

Table 2. Genetic Algorithm parameters.

Population Size	128	Iterations	100k
Crossover Probability	90%	Mutation Probability	10%
Elitism Rate	2%	Tournament Size	9
Forced Edge Failure	10%	Max. Modifiable Bus Routings	10

The results from applying a GA with each crossover operator including ER-Q to the bus scheduling scenarios are shown in Table 3. From these results it can be observed that the quality based crossover operator ER-Q outperforms all other crossover operators in terms of minimising bus fleet distance for all routing scenarios. Of the alternative crossover operators OX achieves the next best results although ER-Q improves upon these non-service distances by as much as 25%. The reduction in electric bus fleet non-service distance by ER-Q is due to the number of buses used in each fleet. ER-Q finds solutions with significantly fewer buses in the optimised solutions. Optimal solutions for routing problems generally minimise vehicle use. A key reason behind ER-Q using fewer electric buses is that through using a quality metric to assess edges, parental or otherwise, returning early to the depot will only occur when close as the edge is of higher quality. Otherwise, further routes are most likely scheduled. The alternative crossover operators all recombine parent solutions introducing new edges without any quality measure meaning that a bus could be inadvertently returned to the depot through the recombination process. Survival of the fittest can be relied upon to some extent to remove these weaker solutions but clearly the dynamics of Darwinian evolution are not enough.

However, the effectiveness of a quality metric within a crossover operator does come with a computational cost due to quality analysis of available edges. The runtimes using each crossover operator are shown in Table 3 whereby it can be observed that ER-Q is the most computationally expensive. For the smaller problem scenarios ER-Q is approximately 2.5x slower than the fastest operator CX although this difference diminishes as the problem size increases. But given the improvement in results this additional computational cost is acceptable as ER-Q is capable of deriving better results within the same time period.

It could be considered that the maximum permissible number of random bus routings that can be modified from parent solutions of ten buses out of up to 900 is too low. The other bus routes are preserved from parent solutions, each parent being responsible for one offspring. For the modifiable bus routings edges

Table 3. Average electric bus fleet not in service distance travelled, buses utilised and evolution runtimes using ER-Q crossover and a range of other crossover operators.

Routing Problem	Crossover Operator	Non-Service Distance Travelled (Km)	Buses Utilised	Runtime (secs)
Scenario A	OX	1529.66 ± 43.39	37.12 ± 1.27	18.07 ± 0.26
	CX	2446.20 ± 83.10	58.36 ± 2.40	17.63 ± 0.20
	PMX	2587.54 ± 85.18	61.08 ± 2.34	22.91 ± 0.48
	ER	2060.09 ± 78.90	49.32 ± 1.91	33.87 ± 0.56
	PX	2351.22 ± 105.12	56.36 ± 2.27	27.24 ± 1.00
	ER-Q	**1401.10 ± 50.89**[†]	33.40 ± 1.15	32.35 ± 0.37
Scenario B	OX	1357.17 ± 45.05	32.64 ± 1.19	16.74 ± 0.13
	CX	2140.56 ± 99.51	52.28 ± 1.93	16.51 ± 0.26
	PMX	2183.24 ± 73.65	52.88 ± 2.07	21.51 ± 0.43
	ER	1987.70 ± 81.52	48.44 ± 2.58	32.17 ± 0.57
	PX	2025.39 ± 79.23	50.40 ± 2.24	25.57 ± 0.71
	ER-Q	**1244.79 ± 52.53**[†]	29.72 ± 1.28	29.97 ± 0.26
Scenario C	OX	5022.58 ± 135.23	146.00 ± 1.66	41.44 ± 0.38
	CX	5316.14 ± 109.33	148.16 ± 1.18	41.27 ± 0.27
	PMX	5257.85 ± 114.20	148.00 ± 1.15	41.93 ± 0.33
	ER	6566.84 ± 386.25	149.28 ± 0.79	73.62 ± 0.50
	PX	5308.84 ± 114.45	147.16 ± 1.91	57.11 ± 1.25
	ER-Q	**4106.50 ± 84.93**[†]	126.80 ± 2.10	74.15 ± 0.44
Scenario D	OX	2462.81 ± 81.06	111.16 ± 2.66	27.40 ± 0.19
	CX	2571.52 ± 101.85	131.08 ± 2.20	27.55 ± 0.20
	PMX	2555.53 ± 90.05	132.40 ± 3.72	27.92 ± 0.45
	ER	2693.05 ± 103.65	125.92 ± 2.97	50.83 ± 0.47
	PX	2543.01 ± 101.07	130.88 ± 3.10	41.95 ± 2.70
	ER-Q	**1976.44 ± 78.27**[†]	94.04 ± 1.84	50.14 ± 0.37
Scenario E	OX	3009.79 ± 83.22	127.04 ± 2.44	33.18 ± 0.17
	CX	3225.10 ± 91.25	135.36 ± 2.31	33.28 ± 0.17
	PMX	3238.47 ± 89.10	136.72 ± 2.11	33.40 ± 0.21
	ER	3528.27 ± 106.10	141.04 ± 3.21	61.10 ± 0.57
	PX	3190.02 ± 72.31	135.52 ± 2.37	48.69 ± 2.11
	ER-Q	**2595.11 ± 41.23**[†]	109.44 ± 1.50	59.72 ± 0.42
Scenario F	OX	1174.93 ± 51.29	114.12 ± 2.77	32.58 ± 0.19
	CX	1319.85 ± 53.19	130.4 ± 3.62	33.04 ± 0.20
	PMX	1338.45 ± 54.81	132.2 ± 2.83	33.69 ± 0.87
	ER	1436.26 ± 80.14	125.92 ± 2.72	60.04 ± 0.53
	PX	1319.18 ± 67.48	129.64 ± 2.89	48.98 ± 2.40
	ER-Q	**903.44 ± 27.04**[†]	87.36 ± 2.27	59.27 ± 0.31

(continued)

Table 3. (*continued*)

Routing Problem	Crossover Operator	Non-Service Distance Travelled (Km)	Buses Utilised	Runtime (secs)
Scenario G	OX	9856.68 ± 127.40	264.28 ± 3.86	92.62 ± 0.29
	CX	11966.71 ± 162.85	322.44 ± 4.54	93.76 ± 0.87
	PMX	11894.88 ± 209.58	328.80 ± 5.48	94.80 ± 0.90
	ER	13124.85 ± 240.66	327.88 ± 6.27	149.77 ± 0.63
	PX	11897.18 ± 126.69	321.20 ± 5.09	127.47 ± 3.01
	ER-Q	**7659.55 ± 148.69**[†]	214.84 ± 3.33	135.82 ± 0.45
Scenario H	OX	7156.48 ± 122.06	344.48 ± 4.85	115.84 ± 0.24
	CX	7802.91 ± 126.00	389.32 ± 6.11	118.54 ± 2.28
	PMX	7637.57 ± 132.83	394.08 ± 4.18	117.92 ± 1.23
	ER	9251.01 ± 396.63	400.40 ± 5.10	182.41 ± 2.67
	PX	7742.30 ± 151.88	388.20 ± 5.25	150.81 ± 2.81
	ER-Q	**5633.01 ± 133.17**[†]	296.12 ± 3.43	163.59 ± 0.44
Scenario I	OX	19854.31 ± 184.29	649.20 ± 9.11	243.56 ± 0.63
	CX	23121.74 ± 248.86	752.44 ± 8.69	247.88 ± 3.55
	PMX	22304.42 ± 275.84	750.00 ± 7.34	254.23 ± 1.84
	ER	26268.26 ± 464.78	772.48 ± 8.39	358.53 ± 0.72
	PX	23044.95 ± 253.68	750.20 ± 7.31	306.16 ± 2.19
	ER-Q	**14432.14 ± 197.34**[†]	536.32 ± 4.12	326.07 ± 1.22

[†] Statistically significant improvement over all other crossover operators with $p < 0.01$, a two-sided significance level and 24 degrees of freedom

from both parents can be utilised as normal with ER-Q crossover. Therefore, the results from using ER-Q crossover are repeated using a range of maximum permissible bus routes that can be modified with the results shown in Fig. 1 in terms of average fleet traversal distance and runtimes. From these results it can be observed that increasing the modification level degrades the derived solutions. Moreover, even using a very small degree of modification is relatively successful. The reason for this is that as the degree of probabilistic decision making increases, the number of potential edges increases, many of which are of lower *quality* but could still be selected. Therefore, it can be hypothesised that ER-Q is a fairly destructive crossover operator and should only make small modifications. Furthermore, using a small degree of modification enables ER-Q to operate faster with little degradation in results.

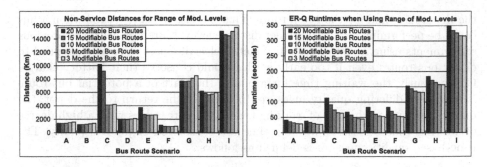

Fig. 1. Average non-service distance travelled and runtimes using ER-Q crossover and a range of permissible bus route modification levels.

5   Defining the Ideal Quality Metric

In the previous experiments, a quality measure was utilised within ER-Q crossover to create a probabilistic model over the available edges to select higher quality edges. Since the goal is to complete all timetabled routes on time in minimal distance, a simple edge length quality metric was deemed sufficient. However, distance is not the only quality measure that could be utilised. Indeed, an issue with using distance between vertices is that it can encourage a bus to return earlier than necessary to the depot. If a bus completes a route finishing close to the depot the distance quality measure of returning to the depot will be high.

Therefore, a question remains over quantifying *quality*. Since a quality measure will operate at a highly localised level, an edge by edge basis, the overall fitness function cannot be utilised. To illustrate the importance of defining quality a range of metrics are proposed. The first considers energy consumption in terms of kWh required by an electric bus to travel out of service to the start of another timetabled bus route. This quality measure is essentially the same as the distance metric. However, a key change is when considering an edge which returns the bus to the depot whereby the edge quality is defined as the energy that could be used to service more bus routes, the remaining charge in the battery. Clearly, if a bus returns to the depot it will no longer be used and any of its remaining energy will be essentially "lost" in terms of servicing the timetable. Thus heuristic information $[\eta_{il}]$ in Eq. (2) for electric bus n can be defined as:

$$[\eta_{il}] = \begin{cases} 1/e_{il} & \text{if vertex } l \text{ is a bus route} \\ 1/c_n & \text{otherwise} \end{cases} \tag{4}$$

where e_{il} is the electric charge required to travel between vertices i and l; c_n is the remaining charge for bus n.

A second quality measure that can be defined for use by ER-Q for electric bus scheduling is the use of time. In the previous experiments due to the detection of constraints on edges a bus is not assigned a route it will arrive late to. However, a bus could arrive early to the beginning of a bus route and then simply wait until the departure time for the given route. This is a loss of service time when a bus could be performing a route a further distance away. Moreover, with the previous experiments using a distance quality metric electric buses were also assigned to closer bus routes even if too early as this would conserve their battery charge. Consequently, the waiting time for a bus at the start of a potential bus route is used as a differing quality measure. In terms of the return to the depot, the quality measure is the loss of working time. Therefore, edges with a high degree of waiting time or loss of working time are considered of poorer quality. The time-based quality measure is defined as follows:

$$[\eta_{il}] = \begin{cases} 1/(s_l - T_n - t_{il}) & \text{if vertex } l \text{ is a bus route} \\ 1/(F - T_n) & \text{otherwise} \end{cases} \tag{5}$$

where s_l is the start time of the route at vertex l; T_n is the current time for bus n; t_{il} is the time to travel between vertices i and l; F is the day finish time.

The time-based quality metric can be extended further. It can be considered that both distance and waiting time are important. Therefore the time metric can be extended to include the travel time between routed along edges. In effect, the total amount of time an electric bus is out of service:

$$[\eta_{il}] = \begin{cases} 1/(s_l - T_n) & \text{if vertex } l \text{ is a bus route} \\ 1/(F - T_n) & \text{otherwise} \end{cases} \tag{6}$$

Finally, two further quality metrics will be used. Firstly, previous results demonstrated that reducing the number of buses used is advantageous. Consequently, a simper quality measure will be used which only labels edges returning to a depot as *poor* quality defined by the loss of service time. Finally, for comparison purposes, no quality measure will be used within ER-Q, in effect a uniform level of quality across edges. The ER-Q process will remain the same using only a small degree of modification of parent bus route schedules. Visibility of the constraints will remain such that any edges that would exceed an electric bus charge level or edges that arrive late to a timetabled route are considered taboo.

The previous experiments are repeated using each individual quality measure with results shown in Fig. 2 in terms of the average not in service distance travelled by the electric bus fleet and the number of buses utilised. A key observation is that all quality measures are not equal, choice of quality measure can significantly influence results. The time-based quality measures perform the best in terms of minimised fleet distance. The number of electric buses used by the time-based quality measures embedded in ER-Q are also lower than any of the other measures. Clearly, any degree of time that a bus is waiting at a bus stop to begin a timetabled route is wasted time that could be put to use. Using a purely distance-based quality measure results in considerable wasted time and hence a greater number of buses are required. This means more trips in and out of the depot increasing the not in service distance travelled by the fleet.

An energy-based quality measure is a slight improvement over a pure distance-based measure. This is due to it being easier to penalise edges that return buses to the depot with charge remaining as observed with a small reduction in buses used. Interestingly, a quality measure that is purely designed to

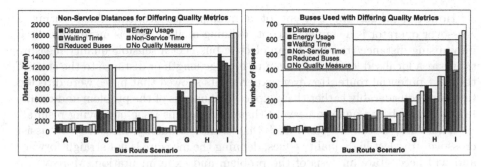

Fig. 2. Average non-service distance travelled and buses utilised for optimised solutions using ER-Q crossover and a range of differing quality measures.

label edges that return a bus to the depot as lower in quality increases bus usage. This is likely due to no time-based quality being used resulting in buses not returning to the depot and assigned a route late in the day wasting much service time. Once the end of the day is reached the constraints dictate that a bus returns to the depot.

Finally, the results in Fig. 2 demonstrate that when using no quality measure, merely problem constraint awareness, the total non-service distance travelled by the fleet is significantly greater than when using distance, energy or time-based quality measures. This reinforces the hypothesis that embedding a measure of edge quality within crossover can improve overall results for permutation routing problems. Moreover, contrasting results in Fig. 2 without using a quality measure to those in Table 3 for the *blind* crossover operators results are still significantly improved. Therefore, even a problem constraint aware crossover operator which is relatively *blind* is beneficial for permutation routing problems.

6   Discussion and Conclusions

This paper has hypothesised that crossover operators for permutation problems operate effectively *blindly* relying on the simple mechanics of Darwinian evolution to improve solutions. These crossover operators have no problem domain knowledge, only that one given solution is better than another. This paper postulated that introducing a degree of problem domain oversight into crossover via the use of an edge quality measure to select parental edges could reduce *blindness* and improve results. In effect, favoring the better aspects of the parents. Hence, a crossover operator, ER-Q, based on Edge Recombination (ER) that probabilistically selects edges based on their perceived quality is used to test this theory. Higher quality edges stand a greater chance of selection. Tested upon a multi-facetted permutation routing problem, the assignment of timetabled bus routes to a fleet of electric buses, the use of a quality-based crossover substantially improved results. Knowledge of the problem domain through a quality measure enables a crossover operator to have oversight of both problem constraints avoiding invalidating solutions and favouring edges beneficial to the problem.

However, whilst the concept of a quality measure embedded within a crossover operator is a simple concept, defining quality itself is more difficult. If minimising distance travelled by an electric bus fleet a sensible quality measure to use at a localised level would seem to be edge length or distance to the start of the next potential route. However, experiments showed that buses were directed to routes considerably earlier than expected. Analysis of the problem and results led to a time-based quality measure that resolved this issue improving results.

Therefore, it can be concluded that whilst a quality measure can assist crossover to select high quality edges, defining the measure is not straightforward and will necessitate analysis of the problem and experimentation. Moreover, a quality measure will most likely be unique to the problem under consideration. This is the advantage of *blind* crossover operators in that they are universally deployable by relying on the dynamics of Darwinian evolution although less effective. Future work will continue to quantify edge *quality* and extending to the detection of low quality edges in parents to earmark them for modification.

Acknowledgements. Supported by Innovate UK [grant no. 10007532] and City Science.

References

1. Ahmed, Z.H.: Genetic algorithm for the traveling salesman problem using sequential constructive crossover operator. Int. J. Biomet. Bioinf. (IJBB) **3**(6), 96 (2010)
2. Applegate, D., Cook, W., Rohe, A.: Chained Lin-Kernighan for large traveling salesman problems. INFORMS J. Comput. **15**(1), 82–92 (2003)
3. Branke, J., Barz, C., Behrens, I.: Ant-based crossover for permutation problems. In: Cantú-Paz, E., et al. (eds.) GECCO 2003. LNCS, vol. 2723, pp. 754–765. Springer, Heidelberg (2003). https://doi.org/10.1007/3-540-45105-6_90
4. Chitty, D.M.: An ant colony optimisation inspired crossover operator for permutation type problems. In: 2021 IEEE Congress on Evolutionary Computation (CEC), pp. 57–64. IEEE (2021)
5. Chitty, D.M.: A partially asynchronous global parallel genetic algorithm. In: Proceedings of the Genetic and Evolutionary Computation Conference Companion, pp. 1771–1778 (2021)
6. Chitty, D.M., Yates, W.B., Keedwell, E.: An edge quality aware crossover operator for application to the capacitated vehicle routing problem. In: Proceedings of the Genetic and Evolutionary Computation Conference Companion, pp. 419–422 (2022)
7. Croes, G.A.: A method for solving traveling-salesman problems. Oper. Res. **6**(6), 791–812 (1958)
8. Dantzig, G.B., Ramser, J.H.: The truck dispatching problem. Manage. Sci. **6**(1), 80–91 (1959)
9. Davis, L.: Applying adaptive algorithms to epistatic domains. In: IJCAI, vol. 85, pp. 162–164 (1985)
10. Dorigo, M., Gambardella, L.M.: Ant colony system: a cooperative learning approach to the traveling salesman problem. IEEE Trans. Evol. Comput. **1**(1), 53–66 (1997)

11. Freisleben, B., Merz, P.: A genetic local search algorithm for solving symmetric and asymmetric traveling salesman problems. In: Proceedings of IEEE International Conference on Evolutionary Computation, pp. 616–621. IEEE (1996)
12. Goldberg, D.E., Lingle, R., et al.: Alleles, loci, and the traveling salesman problem. In: Proceedings of an International Conference on Genetic Algorithms and Their Applications, vol. 154, pp. 154–159. Lawrence Erlbaum Hillsdale, NJ (1985)
13. Grefenstette, J.J.: Incorporating Problem-Specific Knowledge into Genetic Algorithms. Genetic Algorithms and Simulated Annealing, pp. 42–57 (1987)
14. Holland, J.H.: Adaptation in Natural and Artificial Systems: An Introductory Analysis with Applications to Biology, Control, and Artificial Intelligence. U Michigan Press, Ann Arbor (1975)
15. Hu, H., Du, B., Perez, P.: Integrated optimisation of electric bus scheduling and top-up charging at bus stops with fast chargers. In: 2021 IEEE International Intelligent Transportation Systems Conference (ITSC), pp. 2324–2329. IEEE (2021)
16. Janovec, M., Kohani, M.: Grouping genetic algorithm (GGA) for electric bus fleet scheduling. Transp. Res. Procedia 55, 1304–1311 (2021)
17. Kidwai, F.A., Marwah, B.R., Deb, K., Karim, M.R.: A genetic algorithm based bus scheduling model for transit network. In: Proceedings of the Eastern Asia Society for Transportation Studies, vol. 5, pp. 477–489. Citeseer (2005)
18. Kkesy, J., Domański, Z.: Edge recombination with edge sensitivity in TSP problem. Sci. Res. Inst. Math. Comput. Sci. 2(1), 55–60 (2003)
19. Nagata, Y.: Edge assembly crossover: a high-power genetic algorithm for the traveling salesman problem. In: Proceedings of the 7th International Conference on Genetic Algorithms 1997 (1997)
20. Oliver, I., Smith, D., Holland, J.R.: Study of permutation crossover operators on the traveling salesman problem. In: Genetic algorithms and their applications: proceedings of the second International Conference on Genetic Algorithms: July 28–31, 1987 at the Massachusetts Institute of Technology, Cambridge, MA. Hillsdale, NJ: L. Erlbaum Associates (1987)
21. Osaba, E., Carballedo, R., Díaz, F., Perallos, A.: Analysis of the suitability of using blind crossover operators in genetic algorithms for solving routing problems. In: 2013 IEEE 8th International Symposium on Applied Computational Intelligence and Informatics (SACI), pp. 17–22. IEEE (2013)
22. Yiu-Cheung Tang, A., Leung, K.-S.: A modified edge recombination operator for the travelling salesman problem. In: Davidor, Y., Schwefel, H.-P., Männer, R. (eds.) PPSN 1994. LNCS, vol. 866, pp. 180–188. Springer, Heidelberg (1994). https://doi.org/10.1007/3-540-58484-6_262
23. Ting, C.-K.: Improving edge recombination through alternate inheritance and greedy manner. In: Gottlieb, J., Raidl, G.R. (eds.) EvoCOP 2004. LNCS, vol. 3004, pp. 210–219. Springer, Heidelberg (2004). https://doi.org/10.1007/978-3-540-24652-7_21
24. Tinós, R., Whitley, D., Ochoa, G.: A new generalized partition crossover for the traveling salesman problem: tunneling between local optima. Evol. Comput. 28(2), 255–288 (2020)
25. Wang, C., Guo, C., Zuo, X.: Solving multi-depot electric vehicle scheduling problem by column generation and genetic algorithm. Appl. Soft Comput. 112, 107774 (2021)
26. Whitley, D., Hains, D., Howe, A.: Tunneling between optima: partition crossover for the traveling salesman problem. In: Proceedings of the 11th Annual Conference on Genetic and Evolutionary Computation, pp. 915–922 (2009)

27. Whitley, D., Hains, D., Howe, A.: A hybrid genetic algorithm for the traveling salesman problem using generalized partition crossover. In: Schaefer, R., Cotta, C., Kołodziej, J., Rudolph, G. (eds.) PPSN 2010. LNCS, vol. 6238, pp. 566–575. Springer, Heidelberg (2010). https://doi.org/10.1007/978-3-642-15844-5_57

28. Whitley, D.L., Starkweather, T., Fuquay, D.: Scheduling problems and traveling salesmen: the genetic edge recombination operator. In: ICGA, vol. 89, pp. 133–40 (1989)

Maximizing the Number of Satisfied Charging Demands in Electric Vehicle Charging Scheduling Problem

Imene Zaidi[1,2]($\boxtimes$) iD, Ammar Oulamara[2] iD, Lhassane Idoumghar[1] iD,
and Michel Basset[1] iD

[1] Université de Haute-Alsace, IRIMAS UR 7499, 68100 Mulhouse, France
{imene.zaidi,lhassane.idoumghar,michel.basset}@uha.fr
[2] Université de Lorraine, LORIA Laboratory UMR7503, 54506
Vandoeuvre-lès-Nancy, France
ammar.oulamara@loria.fr

Abstract. This paper addresses the electric vehicle charging problem in a charging station with a limited overall power capacity and a fixed number of chargers. Electric vehicle drivers submit their charging demands. Given the limited resources, these charging demands are either accepted or rejected, and an accepted demand must be satisfied. The objective of the scheduler is to maximize the number of satisfied demands. We prove that the problem is NP-hard. Then, we propose a linear programming model, heuristic, and a metaheuristic combining a simulated annealing algorithm with an iterated local search procedure to solve it. We provide computational results to show the efficiency of the proposed methods.

Keywords: Electric Vehicle · Charging Scheduling · Linear Programming · Heuristic · Simulated Annealing

1 Introduction

Electric vehicles have recently gained wide popularity as low-emission vehicles. According to the International Energy Agency [6], the number of electric vehicles reached 16.5 million in 2021. While in 2010, only hundreds of them were on the road. However, the global adoption of electric vehicles is still challenging since charging an electric vehicle is time-consuming and requires considerable electric energy. Moreover, a mass transition to electric vehicles will lead to a saturation of charging stations and a significant increase in electrical power demand that can overload the power grid. Several studies propose smart charging approaches to avoid these negative impacts without expensively upgrading the existing power grid. In smart charging, a management system controls the charging of electric vehicles and optimally schedules the electric vehicle charging load. This paper addresses the electric vehicle charging scheduling problem (EVCSP) in a charging station where drivers submit charging demand reservations before arriving.

Given the lack of charging stations, the short range of electric vehicles, and the long time required to charge them, drivers of electric vehicles need to carefully plan their trips to ensure that they will have opportunities to recharge their batteries. As a result, it is preferable for them to confirm in advance that the charger they intend to use is available. Moreover, the Open Charge Point Protocol includes the reservation functionality of charging stations [1].

The remainder of this paper is organized as follows. Section 2 briefly reviews the main works on EVCSP. Section 3 describes in detail the investigated problem. Section 4 provides the complexity of the problem. Section 5 formulates it as an integer linear programming (ILP) model. The proposed heuristic is presented in Sect. 6. Section 7 details the developed metaheuristic that combines a simulated annealing algorithm with a iterated local search procedure. Section 8 evaluates the performance of the proposed methods. The paper closes with some conclusions and future research directions in Sect. 9.

2   Related Work

We focus on studies that investigated the problem of optimizing the charging load of electric vehicles from the perspective of charging station operators. The main objectives of these operators are to reduce the total charging cost [5, 14, 15] or to maximize the satisfaction of their customers. In smart charging stations, a control system builds a charging schedule while considering the arrival and departure times and the amount of energy requested by each vehicle driver. Many studies assume an uncertain arrival time [4, 13, 15]. Authors in [14] consider that electric vehicles may arrive with or without a reservation. The electric vehicle drivers can provide the departure times [4, 14, 16], or they can be estimated based on historical behavior [15]. As for the desired energy, [14] assume that the electric vehicle drivers directly specify their desired energy in kWh. Other papers consider charging electric vehicles to the rated battery capacity [10, 12, 16]. For constraints related to the charging station, authors in [10, 12, 16] consider a variable charging power where the charging rate varies over time, while in [4], constant power rates were considered. One of the most commonly used constraints is the capacity of the charging infrastructure. This constraint defines the total power limit of the charging infrastructure, expressed in (kW). Limiting the total charging load of electric vehicles is essential to keep the power peaks low and avoid overloading other equipment and transmission lines. Different optimization approaches were adapted and developed to solve EVCSP. A two-stage approximate dynamic programming was proposed in [16]. Some studies have considered stochastic optimization methods as in [13], where the authors proposed a stochastic linear programming model to schedule the electric vehicles charging load in real-time. Metaheuristics were also applied to solve the EVCSP. For example, we can find a particle swarm optimization in [12, 14], a genetic algorithm in [5], a GRASP-like algorithm and a memetic algorithm in [4]. Although the studies mentioned above have examined various aspects of the EVCSP, the charging station operating model, the constraints, and the optimization objective are different from this study. Thus, comparing results between the proposed methods and literature cannot be pertinent.

3 Problem Description

The formulation of an instance of the EVCSP can be defined as follows. We have a set $\mathcal{J} = \{1, \ldots, n\}$ of charging demands to be scheduled on a set of $\mathcal{M} = \{1, \ldots, m\}$ of chargers. Each charger i delivers a constant power of w_i (kW). The total power that can be delivered by all chargers simultaneously must not exceed w_G (kW), which will further be denoted as the power grid capacity. Each electric vehicle j has an arrival time r_j, departure time d_j, and an energy requirement e_j (kWh) that must be satisfied by its departure time d_j. The charging time p_{ij} of each demand j when assigned to charger i is equal to $p_{ij} = \frac{e_j}{w_i}$. Charging demands can either be accepted or rejected. When a charging demand is accepted, it must be satisfied. At each time, a charger can only charge one vehicle, and a vehicle can only be charged by one charger. During the time interval $[r_j, d_j)$, the vehicle is parked and plugged into charger i. The charging scheduling is preemptive, i.e., the charging operation of each vehicle j can be interrupted at any time and resumed later in the interval $[r_j, d_j)$. Even when the vehicle completes charging before d_j, it still occupies the charger i until it departs. Unless otherwise mentioned, we divide the scheduling time horizon H into T time slots of equal length τ. The scheduling objective is to maximize the number of satisfied charging demands.

4 Complexity

Theorem 1. *The problem of maximizing the number of satisfied charging demands is NP-hard.*

Proof. We show that the problem is NP-hard by proving that its special case where all chargers are identical is NP-hard. Let $\overline{m} = \lfloor \frac{w_G}{w} \rfloor$, where w is the charging power rate of each charger. Clearly, at each time, at most $\overline{m}$ chargers can be activated at the same time. Furthermore, maximizing the number of satisfied charging demands is equivalent to minimizing the number of rejected demands. Minimizing the number of rejected charging demands is equivalent to minimizing the number of late jobs in $\overline{m}$ identical parallel machines scheduling problem with release date and preemption of jobs ($P_{\overline{m}}|prmt| \sum U$). In scheduling problem $P_{\overline{m}}|prmpt| \sum U$ if a job is late in an optimal schedule, it is immaterial where it is scheduled. Thus, scheduling on-time jobs is important. In the optimal schedule, the on-time jobs are scheduled in their interval $[r_j, d_j]$, and at most $\overline{m}$ are used to schedule these jobs. Then the on-time jobs correspond to the set of accepted demands in the problem of maximizing the number of satisfied charging demands. In [3] authors showed that the problem $P_{\overline{m}}|prmpt| \sum U$ is NP-Hard even with two identical machines. Then the problem of maximizing the number of satisfied demands is NP-Hard.   □

5   Mathematical Formulations

In this section, we formulate the described problem as an integer linear programming (ILP) model. We define binary variables s_{ij} to specify whether or not the charging demand of electric vehicle j is scheduled on charger i. In addition, we define binary variables x_{ijt} specifying whether or not the electric vehicle j is plugged into the charger i at time slot t. Also, we introduce binary variables y_{jt} that specifies whether or not the electric vehicle j is charging at time slot t. The mathematical formulation is as follows.

$$\max \quad \sum_{j=1}^{n} \sum_{i=1}^{m} s_{ij} \tag{1}$$

$$\sum_{i=1}^{m} s_{ij} \leq 1 \qquad \forall j \in \mathcal{J} \tag{2}$$

$$\sum_{j=1}^{n} x_{ijt} \leq 1 \qquad \forall i \in \mathcal{M}, \quad t \in H \tag{3}$$

$$\sum_{t=r_j}^{d_j} x_{ijt} = s_{ij}(d_j - r_j) \qquad \forall i \in \mathcal{M}, \quad j \in \mathcal{J} \tag{4}$$

$$\sum_{t=r_j}^{d_j} y_{jt} = \sum_{i=1}^{m} p_{ij} s_{ij} \qquad \forall j \in \mathcal{J} \tag{5}$$

$$\sum_{j=1}^{n} \sum_{i=1}^{m} w_i \times s_{ij} \times y_{jt} \leq w_G \qquad \forall t \in H \tag{6}$$

Constraints (2) ensure that when a demand j is accepted, it is assigned to one charger. Constraints (3) ensure that each charger i charges one demand at each time slot t. Constraints (4) ensure that if a charging demand j is accepted to be scheduled on charger i, then it will be plugged into this charger from its arrival r_j to its departure d_j. Constraints (5) ensure that if a charging demand j is accepted, the vehicle j will be charged to its requested energy. Constraints (6) ensure that at each time slot t, the total power delivered by all chargers does not exceed w_G. In addition, variables x_{ijt} and y_{jt} are set to zero for all t where $t < r_j$ and $t \geq d_j$.

Constraints (6) can be linearized by using a new binary variables z_{ijt}. a variable z_{ijt} equals to 1 if variables y_{jt} and s_{ij} equal to 1. Constraints (6) are replaced by the following constraints:

$$z_{ijt} \geq y_{jt} + s_{ij} - 1 \qquad \forall i \in \mathcal{M}, \quad j \in \mathcal{J}, \quad t \in H \tag{7}$$

$$z_{ijt} \leq y_{jt} \qquad \forall i \in \mathcal{M}, \quad j \in \mathcal{J}, \quad t \in H \tag{8}$$

$$z_{ijt} \leq s_{ij} \qquad \forall i \in \mathcal{M}, \quad j \in \mathcal{J}, \quad t \in H \tag{9}$$

$$\sum_{i=1}^{k} \sum_{j=1}^{n} w_i z_{ijt} \leq w_G \qquad t \in H \tag{10}$$

6   Greedy Constructive Heuristic

Since maximizing the number of satisfied demands is NP-Hard, it is hard to find optimal solutions for large-size instances in a reasonable time. Moreover, using a commercial linear programming solver may incur additional costs for charging station operators. Hence, we propose heuristics and metaheuristics. The proposed heuristic, detailed in Algorithm 1, builds a charging schedule by considering vehicles in the non-decreasing order of their arrival time r_j and breaking ties first by the non-decreasing order of their departure time d_j, then by the non-decreasing order of their energy request e_j. Let $(w_G^t)_{t \in H}$ be a vector of reals that stores the power allocated at each time slot t, which is initialized to 0. For each vehicle j, if at least a charger is available at r_j, the heuristic begins by seeking an available charger with the smallest charging power to charge j without exceeding the current grid capacity (lines 7-10). If such a charger exists, it is selected to charge vehicle j (line 11). Otherwise, the heuristic calculates the value $e(j', r_j, d_j)$ that represents the amount of energy allocated to each scheduled charging demand j' ($j' \neq j$) in the interval $[r_j, d_j)$. The charging demand with the greatest value of $e(j', r_j, d_j)$ will be rejected if $e(j', r_j, d_j)$ is greater than the requested energy e_j (line 15). Otherwise, the vehicle j is rejected (line 16). When no charger is available at r_j (lines 18-22), the charging demand with the maximum departure time is rejected.

Algorithm 1: Constructive greedy heuristic

Input : The set of charging demands $\mathcal{J}$, the set of chargers $\mathcal{M}$, the grid capacity w_G

Output: The assignment of vehicles to chargers, the set of rejected demand

1 Sort $\mathcal{J}$ by non-decreasing order of r_j. Then, in non-decreasing order of d_j. Then, in non-decreasing order of e_j ;

2 Sort $\mathcal{M}$ by non-decreasing order of charging power w_i ;

3 $(w_G^t) \leftarrow (0)_{t \in H}$;

4 **while** $\mathcal{J} \neq \emptyset$ **do**

5   Let j be the first demand in $\mathcal{J}$;

6   **if** *at least a charger is available at r_j* **then**

7   w_j^a be the first available charger in $\mathcal{M}$;

8   Let b be the number of time slots in $[r_j, d_j)$ where $w_G^t + w_j^a \leq w_G$;

9          $E_j \leftarrow e_j/(b \times \tau)$;

10   **if** *the vehicle j can be scheduled on an available charger i with a charging power $w_i \geq w_j$ without exceeding w_G* **then**

11   Schedule j on charger i and remove it from $\mathcal{J}$;

12   **else**

13   Let $e(j', r_j, d_j)$ be the allocated energy to charging demand $j' \neq j$ in the interval $[r_j, d_j)$;

14   Let k be the scheduled demand with $\max_{j'} e(j', r_j, d_j)$;

15   **if** $e(k, r_j, d_j) > e_j$ **then** Reject k ;

16   **else** Reject j and remove it from $\mathcal{J}$

17   **end**

18   **else**

19   Let j' be the scheduled charging demand with maximum $d_{j'}$;

20   **if** $d_{j'} > d_j$ **then** Reject j' ;

21   **else** Reject j and remove it from $\mathcal{J}$

22   **end**

23   Update w_G^t;

24 **end**

7   Simulated Annealing Metaheuristic

7.1   Solution Representation

Solving the scheduling problem consists of two main decisions: first, selecting electric vehicles to be plugged into chargers; then, selecting vehicles to charge by choosing the appropriate time slots for charging without exceeding w_G. Therefore, a solution to the charging scheduling problem consists of the assignment solution and the power allocation solution. We represent the assignment of charging demands to chargers as a vector $(\pi_1, .., \pi_m)$ where π_i is the sequence of vehicles assigned to a charger i and we place the unassigned demands in a list of rejected demands. The power allocation solution is represented with a vector $(T_j)_{j \in \mathcal{J}}$ where $T_j \subseteq H$ stores, for each vehicle j, the time slots chosen for charging process. For convenience, we define the vector $(w_G^t)_{t \in H}$, which stores the minimum grid capacity at each time slot t.

7.2   Simulated Annealing

The simulated annealing (SA) algorithm, initially proposed by [7], is a stochastic local search metaheuristic successfully adapted to address several scheduling problems. In this paper, a candidate solution for the SA algorithm represents the assignment of charging demands to chargers, on which an iterated local search (ILS), described in Sect. 7.4, is applied to get the power allocation vector and the objective function value of each generated solution. The detailed procedure of the implemented SA is presented in Algorithm 2. It starts by taking as input an initial solution (S_0), generated using the heuristic detailed in Sect. 6, and five parameters: the maximum number of generated neighbors at each iteration ($maxGenerated$), the acceptation ratio at each iteration ($acceptanceRatio$), the final temperature (T_f), the maximum global number of generated solutions ($maxTrials$), and the parameter for initializing the value of the temperature (μ). First, the initial solution S_0 is set as the current solution S and as the global best solution S_{best} (line 1). The temperature parameter T is initially set to a value proportional to the objective function value of the initial solution $T = \mu f(S_0)$. The maximum number of accepted solutions at each iteration ($maxAccepted$) is initially set proportionally to the parameter ($maxGenerated$) (line 2). At each iteration (lines 3-19), SA generates neighbors of the current solution S until reaching either ($maxGenerated$) or ($maxAccepted$). We detail the neighborhood generation in Sect. 7.3. For each new solution S', the global number of generated solutions ($trial$) and the number of generated neighbors of S' ($generated$) are incremented (lines 8-9). The objective function value of each solution, i.e., the number of scheduled demands, is referred to by $f(S)$, and it is calculated by the ILS procedure given in Sect. 7.4. The gap between the objective values of the new solution S' and the current solution S is calculated as $\Delta f = f(S') - f(S)$. The neighbor S' is accepted and replaces the current solution based on the Metropolis criteria (lines 10-16); the new solution S' replaces the current solution if there is an improvement, i.e., $\Delta f > 0$. If S' improves the best solution found so far, it will become the new global best solution S_{best}. Otherwise, a random number is generated following the uniform distribution $U[0, 1]$ and the neighbor S' will become the current solution if $U(0, 1) \leq e^{\Delta f/T}$ where T is the temperature parameter that controls the probability of accepting worse solutions. For each accepted solution, the parameter $accepted$ is incremented (line 12). Finally, a cooling scheme gradually decreases the temperature at each iteration (line 16). We consider the Lundy-Mees cooling scheme proposed by [9]. It updates the temperature T at each iteration l as $T_{l+1} = \frac{T_l}{a+bT_l}$. Connolly in [2] develops a variant of the Lundy-Mees scheme that set the parameter a to 1 and b in function of the initial temperature T_0, the final temperature T_f and the size of the neighborhood M as $b = \frac{T_0 - T_f}{MT_0 T_f}$. Here, the number of iterations is not fixed directly. In fact, if we omit the condition on $maxAccepted$, the number of iterations will be equal to $\frac{maxTrials}{maxGenerated}$. Thus, we set M to this value (line 1). After updating the temperature, the number of generated neighbors ($generated$) and the number of accepted solutions ($accepted$) are reset to zero (line 4). The algorithm will stop if the number of generated solutions ($trial$) reaches its maximum ($maxTrials$), or after generating ($maxGenerated$) solutions that did not result in accepted solutions, i.e. $accepted = 0$ (line 17). When the stopping criterion is met, the algorithm terminates and returns the best solution S_{best} found so far.

Algorithm 2: Simulated annealing

    input : S_0, $maxGenerated$, $acceptanceRatio$, T_f, $maxTrials$, μ
    output: Best solution found S_{best}

1 $S_{best} \leftarrow S_0$, $S \leftarrow S_0$, $T \leftarrow \mu f(S_0)$, $M \leftarrow \frac{maxTrials}{maxGenerated}$, $trial \leftarrow 0$;

2 $maxAccepted \leftarrow acceptanceRatio \times maxGenerated$;

3 **repeat**

4     $accepted \leftarrow 0$; $generated \leftarrow 0$;

5   **while** $generated \leq maxGenerated$ **and** $accepted \leq maxAccepted$ **do**

6         $S' \leftarrow Neighbor(S)$;

7         $\Delta f \leftarrow f(S') - f(S)$;

8         $generated \leftarrow generated + 1$;

9         $trial \leftarrow trial + 1$;

10   **if** $\Delta f > 0$ **or** $U(0, 1) \leq e^{\Delta f/T}$ **then**

11           $S \leftarrow S'$;

12           $accepted \leftarrow accepted + 1$;

13   **if** $f(S) > f(S_{best})$ **then** $S_{best} \leftarrow S$;

14   **end**

15   **end**

16     $T \leftarrow \frac{T}{1+bT}$;

17 **until** $trial \leq maxTrials$ **and** $accepted > 0$;

18 **return** S_{best}

7.3   Neighborhood Operators

The SA algorithm randomly chooses one of three operators to generate a new solution:

- **Change assignment:** this operator chooses a charging demand j on a charger i_1 and moves it to another charger i_2. The chargers and the charging demand are randomly selected. If a charging demand in charger i_2 overlaps with j, the move is discarded.
- **Assign a rejected charging demand:** this operator chooses a charging demand j from the rejected list and inserts it on a charger i. The charger and the charging demand in the rejected list are randomly selected. The move is discarded if at least one charging demand in charger i overlaps with j.
- **Reject a charging demand:** this operator moves a charging demand from a charger to the rejected list. The charger and the charging demand are randomly selected.

When a move is discarded, the SA algorithm randomly selects another operator. After each successful move, the SA algorithm applies an ILS procedure to construct and improve the power allocation solution.

7.4   Iterated Local Search

Given an assignment solution, the iterated local search (ILS) procedure will solve the power allocation problem by selecting the maximum subset of scheduled demands from the assignment solution that can be satisfied without exceeding the grid capacity w_G. The assignment solution may or not be feasible, i.e., the grid capacity w_G may not be sufficient. Let $\mathcal{J}'$ be the set of assigned charging demands. Let $\tilde{w}_G$ be the minimum grid capacity required to satisfy all charging demands in the set $\mathcal{J}'$. The basic idea is to obtain a charging schedule with the minimum value of $\tilde{w}_G$. When $\tilde{w}_G > w_G$, we insert and reject charging demands until $\tilde{w}_G$ reaches w_G. Note that we can only insert the demands rejected by the ILS procedure. Moreover, each demand can only be reinserted in their previously assigned charger, meaning that we cannot move a charging demand to another charger. Therefore, we keep a list L_{LS} of rejected demands by ILS along with their previous chargers. The implemented ILS algorithm (Algorithm 3) starts by building the power allocation vectors of an assignment solution S_0 using a heuristic described in Algorithm 4 (line 1). The current solution S is set to S_0. At each iteration (line 3-23), the ILS procedure generates $maxGeneratedLS$ neighbors of the current solution S (line 5-12). For each generated neighbor, it applies a procedure to minimize $\tilde{w}_G$ (line 7) that will be described below. The best feasible solution S' in the neighborhood of S is selected. A solution is feasible if its grid capacity $\tilde{w}_G$ is less than or equal to w_G. If the best neighborhood S' is better than the best solution found so far S^*, it will replace the current solution S and the best solution S^*. Otherwise, the number of non-improving iterations $iter$ is incremented (line 21). In this case, the current solution S is set to either the best solution in the neighborhood S' or to S^*. The best solution in the neighborhood S' may replace the current solution S if a randomly generated number u is less than the probability p_{iter} (line 17-22). p_{iter} decreases in a geometric way [11] and is calculated as follows. $p_{iter} = p_0 \times r^{iter-1}$, Where p_0 is the initial acceptance probability, $r < 1$ is the reducing factor, and $iter$ is the number of iterations. When the number of non-improving iterations $iter$ exceeds $maxNonImproving$, the search is considered as stagnating on a local optimum and is subsequently terminated.

Algorithm 3: Iterated local search procedure

Input : The assignment solution S_0, $maxNonImproving$, $maxGeneratedLS$, r, p_0
Output: Best feasible solution found S^*
1 Initialize the power allocation for S_0 according to Algorithm 4;
2 $iter \leftarrow 0$; $S \leftarrow S_0$; $S^* \leftarrow$ empty solution;
3 **while** $iter < maxNonImproving$ **do**
4     $S' \leftarrow$ empty solution;
5   **for** $k = 1$ *to* $maxGeneratedLS$ **do**
6        $N \leftarrow$ Local Neighbor(S);
7   Apply minimizing grid capacity procedure to N;
8   **if** $\tilde{w}_G(N) \leq w_G$ and $f(N) > f(S')$ **then**
9   **if** S^* *is empty* **then** $S^* \leftarrow N$;
10           $S' \leftarrow N$;
11   **end**
12   **end**
13   **if** $f(S') > f(S^*)$ **then**
14        $S \leftarrow S'$;
15        $S^* \leftarrow S'$;
16        $iter \leftarrow 0$;
17   **else**
18   Generate a random number $u \sim U(0, 1)$;
19   **if** $u < p_0 \times r^{iter-1}$ **then** $S \leftarrow S'$;
20   **else** $S \leftarrow S^*$;
21        $iter \leftarrow iter + 1$
22   **end**
23 **end**
24 **return** S^*

Initial Solution for Power Allocation. Let $\mathcal{J}'$ be the set of vehicles in the assignment solution vector. Let w_j be the charging power of each vehicle $j \in \mathcal{J}'$. Then, the charging time p_j of each demand j can be calculated as $\lceil e_j/w_j \rceil$. The proposed heuristic, detailed in Algorithm 4, builds the power allocation solution for the set $\mathcal{J}'$ by considering the assigned vehicles in the non-decreasing order of their departure time d_j, and break ties first by non-increasing order of their energy request e_j, then by non-increasing order of their arrival time r_j (line 1). The grid capacity $\tilde{w}_G$ and power allocation vectors are initialized to 0 (line 2). The power allocation heuristic starts by charging vehicle j at time slots without exceeding $\tilde{w}_G$ in chronological order (lines 4, 6–15). Then, on time slots with the minimum w_G^t value (lines 5, 6–15).

Algorithm 4: Power allocation heuristic

Input : The set of charging demands $\mathcal{J}'$, the selected charging power w_j and the charging time p_j for each vehicle

Output: The vectors $(T_j)_{j \in \mathcal{J}}$, $(w_G^t)_{t \in H}$, the grid capacity $\tilde{w}_G$

1 Sort $\mathcal{J}'$ by non-decreasing order of d_j. Then, in non-increasing order of e_j. Then, in non-increasing order of r_j ;

2 $(w_G^t) \leftarrow (0)_{t \in H}$; $\tilde{w}_G \leftarrow 0$; $(T_j) \leftarrow (\emptyset)_{j \in \mathcal{J}'}$;

3 **for** $j \in \mathcal{J}'$ **do**

4 $\quad$ $H_1 \leftarrow$ the set of time slots t where $t \in [r_j, d_j)$ and $\tilde{w}_G \geq w_j + w_G^t$ sorted in chronological order;

5 $\quad$ $H_2 \leftarrow$ the set of time slots t where $t \in [r_j, d_j)$ and $t \notin H_1$ sorted in non decreasing order of w_G^t;

6 $\quad$ **while** $p_j > 0$ **do**

7 $\quad\quad$ **if** $H_1 \neq \emptyset$ **then** $H_i \leftarrow H_1$;

8 $\quad\quad$ **else** $H_i \leftarrow H_2$;

9 $\quad\quad$ Let t be the first time slot of H_i;

10 $\quad\quad$ Remove t from H_i ;

11 $\quad\quad$ $T_j \leftarrow T_j \cup \{t\}$;

12 $\quad\quad$ $w_G^t \leftarrow w_G^t + w_j$;

13 $\quad\quad$ $p_j \leftarrow p_j - 1$;

14 $\quad\quad$ **if** $w_G^t > \tilde{w}_G$ **then** $\tilde{w}_G \leftarrow w_G^t$

15 $\quad$ **end**

16 **end**

17 **return** $\tilde{w}_G$, $(T_j)_{j \in \mathcal{J}}$, $(w_G^t)_{t \in H}$

Local Neighbor Structure. In the ILS procedure, a neighbor is generated by one of the following operators:

- **Reject** this operator removes one or multiple charging demands from a charger to the rejected list L_{LS}. We implements three methods to select a vehicle to reject. First, a randomly chosen vehicle. Second, reject the vehicle j with the greatest value v_j where $v_j = \sum_{t \in T'} w_G^t$ where $T' = \{t \in T_j$ and $w_G^t > w_G\}$. Third, calculate the value v_j for all scheduled vehicles and then a roulette wheel selection [8] is performed i.e., a vehicle j with a higher value v_j has a higher probability to be chosen. After rejecting a vehicle, the w_G^t is updated.
- **Reinsert** this operator randomly chooses one or more vehicles from L_{LS} to be assigned back to its charger. The power allocation for inserted vehicle is obtained using the same procedure in Algorithm 4 (lines 4-21).

Minimizing Grid Capacity Procedure. We use a SA algorithm similar to Algorithm 2 but with different objective function $f(S)$ and different neighbor structure. This second SA is denoted by MINWG-SA. The objective of MINWG-SA is to try to reschedule the charging operations so that the minimum grid capacity $\tilde{w}_G$ is minimized. Since Algorithm 2 is a maximization algorithm, in the MINWG-SA, we replace line 10 by $\Delta f < 0$ or $U(0,1) \leq e^{-\Delta f/T}$. Also, line 13 is replaced by $f(S) < f(S_{best})$. A neighbor structure in the minimizing grid capacity local search method moves a charging operation of a vehicle j from a time slot $t_1 \in T_j$ to another time slot $t_2 \notin T_j$. Let $\mathcal{J}'$ be the set of scheduled charging demands where $d_j - r_j - p_{ij} > 0$, where p_{ij} is the charging time of vehicle j on its assigned charger type i. First, we randomly select an electric vehicle $j \in \mathcal{J}'$ and two time slots t_1 and t_2, where t_1 is a time slot with $w_G^{t_1} = \min_{\{t \in T_j\}} w_G^t$, and t_2 is a time slot with $w_G^{t_2} = \max_{\{t \notin T_j\}} w_G^t$. Then, the charging operation of vehicle j is moved from time t_1 to t_2 by deleting t_1 from T_j and adding t_2 to T_j. This procedure is repeated k times for the same vehicle, where k is randomly selected in $\{1, \ldots, p_{ij}\}$. After each move, the vector w_G^t is updated as well as the objective value $\tilde{w}_G$.

8 Simulation Results

The proposed algorithms are implemented in C++, and run on a desktop computer with an Intel Core i5, 2.9 GHZ CPU and 8 GB RAM. The ILP model is solved using CPLEX 12.8. In the following, we present our experimental results on randomly generated instances.

We consider five groups of instances with different number of charging demands $n \in \{10, 40, 50, 100\}$, different number of chargers $m \in \{15, 24, 27, 30\}$, and different power grid capacities $w_G \in \{50, 75, 100, 125\}$. For each group, one third of chargers deliver a power $w_1 = 11(\text{kW})$, one third of chargers deliver a power $w_2 = 22$ (kW), and the remaining third of chargers deliver a power $w_3 = 43$ (kW). For each group, we generate 10 different random instances as follows. The arrival times of vehicles are generated from the uniform distribution in the interval $[0, 0.2n]$ (in hours). The required energy are generated from the uniform distribution $[5.5, 66]$ (in kWh). To generate the departure times of vehicles, we first calculate the charging times p_{1j} (in hours) for each vehicle j $\in \mathcal{J}$ assuming that it can be charged with chargers of type 1 (11 kW). Then, the departure time of each vehicle j is calculated as $d_j = r_j + (1 + \alpha)p_{1j}$, where α is randomly chosen according to the value p_{1j} as follows. For p_{1j} in $[0.5, 1]$, $[1, 2]$, $[2,3]$, $[3,4]$, $[4,5]$, and $[5,6]$ α is randomly chosen in $[0.1, 1]$, $[0.1, 0.9]$, $[0.1, 0.8]$, $[0.1, 0.7]$, $[0.1, 0.6]$, and $[0.1, 0.5]$, respectively. On the basis of preliminary experiments, we set the parameters μ, $maxGenerated$, $maxTrials$, $acceptanceRatio$, and T_f to 0.12, 50, 100, 0.1, and 0.001 respectively. For the LS procedure, we set the parameters $maxNonImproving$, $maxGeneratedLS$, the reducing factor r and the initial acceptance p_0 to 5, 5, 0.75, and 0.2 respectively.

We set the maximum computation time of CPLEX to 30 min. Table 1 provides a comparison of results obtained for the four groups of instances. The first column denotes the instance number in the group. For CPLEX and the heuristic, column "scheduled" reports the objective value found, and column "time" displays the total running time in seconds. Due to the stochastic nature of the SA algorithm, ten independent executions were done for each instance. We report the best, the worst, and the average objective function value over the ten runs in columns "best", "worst", and "average", respectively. Also, we report the standard deviation of the mean objective function value in column "std" and the average running time in column "time".

Table 1. Comparison results between CPLEX, the heuristic, and the SA algorithm.

instance	CPLEX		Heuristic		SA				
	scheduled	time (s)	scheduled	time (s)	best	worst	average	std	time (s)
group 1 with $n = 10$, $m = 15$, and $w_G = 50$									
1	10	6.07	7	5.45E-05	10	10	10	0.00	3.56
2	9	1800.66	7	3.89E-05	9	9	9	0.00	23.12
3	9	1800.38	7	2.67E-05	9	9	9	0.00	19.54
4	10	3.98	7	8.70E-05	10	9	9.4	0.52	17.49
5	9	537.35	9	1.96E-05	9	9	9	0.00	21.66
6	10	18.17	6	4.74E-05	10	8	9.6	0.70	12.94
7	9	1800.48	7	3.17E-05	9	8	8.3	0.48	24.85
8	9	742.27	7	3.50E-05	9	8	8.7	0.48	24.43
9	9	46.03	6	5.14E-05	9	8	8.1	0.32	22.92
10	9	1800.66	7	3.21E-05	9	8	8.6	0.52	23.82
Average	9.30	855.60	7.00	4.24E-05	9.30	8.60	8.97	0.30	19.43
group 2 with $n = 40$, $m = 24$, and $w_G = 75$									
1	26	1802.38	27	1.37E-04	30	29	29.1	0.32	27.60
2	26	1802.27	25	2.19E-04	31	29	29.6	0.70	26.04
3	11	1801.35	23	2.08E-04	26	24	25.2	0.63	29.64
4	23	1802.33	19	4.86E-04	24	22	23.3	0.67	31.95
5	25	1801.64	21	2.17E-04	26	24	25.1	0.74	30.58
6	26	1802.17	25	1.34E-04	29	28	28.7	0.48	29.68
7	26	1801.81	26	1.02E-04	30	28	29.7	0.67	29.79
8	24	1801.51	23	1.66E-04	28	27	27.3	0.48	29.87
9	25	1801.89	23	1.38E-04	27	26	26.6	0.52	32.25
10	30	1802.85	30	9.83E-05	33	31	32.1	0.57	28.48
Average	24.20	1802.02	24.20	1.91E-04	28.40	26.80	27.67	0.58	29.59
group 3 with $n = 50$, $m = 27$, and $w_G = 100$									
1	21	1802.28	36	1.88E-04	41	39	40.20	0.79	34.45
2	31	1802.03	37	1.16E-04	44	42	42.70	0.67	34.62
3	16	1801.76	35	1.06E-04	41	38	39.50	0.85	32.63
4	34	1802.02	38	1.45E-04	44	42	42.80	0.63	30.70
5	22	1802.04	41	1.35E-04	45	43	44.10	0.57	33.17
6	24	1801.96	38	1.69E-04	41	40	40.80	0.42	34.59
7	23	1801.92	37	1.99E-04	42	40	41.10	0.74	32.79
8	31	1801.90	34	4.25E-04	39	37	38.00	0.67	35.74
9	33	1801.90	34	1.53E-04	37	34	36.10	0.88	35.74
10	15	1802.46	40	1.71E-04	45	43	44.00	0.67	33.12
Average	25	1802.03	37	1.81E-04	41.9	39.8	40.93	0.69	33.75

(*continued*)

Table 1. (*continued*)

instance	CPLEX		Heuristic		SA				
	scheduled	time (s)	scheduled	time (s)	best	worst	average	std	time (s)
group 4 with $n = 100$, $m = 30$, and $w_G = 125$									
1	18	1806.49	76	6.08E-04	77	76	76.20	0.42	50.57
2	11	1806.12	81	2.39E-03	86	83	84.90	0.88	49.28
3	13	1805.92	76	2.32E-04	80	78	78.90	0.74	55.75
4	14	1806.03	75	3.54E-04	78	76	77.40	0.70	55.62
5	14	1806.01	75	6.24E-04	82	78	80.50	1.18	52.36
6	13	1805.99	77	2.65E-04	82	79	81.00	0.94	55.05
7	13	1805.57	78	2.43E-04	83	81	81.40	0.70	51.40
8	14	1806.07	73	6.58E-04	80	77	78.70	1.16	50.72
9	12	1806.21	74	3.89E-04	82	80	80.40	0.84	51.84
10	14	1806.14	77	3.88E-04	81	77	79.40	1.26	55.16
Average	13.60	1806.06	76.20	6.15E-04	81.10	78.50	79.88	0.88	52.78

First, we can notice that CPLEX found six optimal solutions out of 40, all in group one instances with ten vehicles (instances 1,4, 5, 6, 8, and 9 in group 1). All remaining instances were hard to solve for CPLEX within 30 min. The SA also achieved six optimal solutions. However, it took an average time of 17.16 s, while CPLEX took an average time of 225.64 s. As expected, the SA algorithm outperforms the heuristic since it is set to the initial solution for the SA algorithm. The SA algorithm achieved the best solutions in all instances. We calculate the average gap $Gap_{best}(\%)$ (resp. $Gap_{mean}(\%)$) between the objective values found by CPLEX S_{CPLEX} and the best (resp. mean) objective values found by the SA algorithm S_{SA} as $Gap_{best}(\%) = \frac{S_{CPLEX} - S_{SA}}{S_{SA}}$. The $Gap_{best}(\%)$ values were 0%, -14.75%, -40.33%, and -83.23% for groups 1, 2, 3, and 4, respectively. The $Gap_{mean}(\%)$ values were 3.68%, -12.54%, -38.92%, and -82.97% for groups 1, 2, 3, and 4, respectively. The gap between the SA algorithm and solutions found by CPLEX increases significantly with the size of instances. The heuristic starts performing better than CPLEX in groups 3 and 4. Finally, we compare the proposed methods in terms of running time. As expected, the heuristic is faster than the SA algorithm. The heuristic took less than one millisecond, whereas the SA algorithm took an average running time of 32.44 s. In summary, the SA algorithm outperformed CPLEX in significantly less time.

9 Conclusion

This paper addressed the EVCSP in a charging station with different charging types and limited overall power. We proved that the problem is NP-Hard and we formulate it as an ILP model. It was hard for CPLEX to solve the ILP model within 30 min. Therefore, we designed a heuristic and a SA algorithm combined with an ILS procedure. We generated different instances to evaluate the performance of the proposed methods. The experimental results underline the efficiency of the proposed methods. We assumed that the data related to vehicle charging demands were known in advance. In future research, we can study the scheduling problem in real-time to handle charging demands with or without reservations. Another challenge is considering multi-objective optimization to add the objective of minimizing the charging costs.

References

1. Ocpp 2.0 - 2.0.1 specification. Tech. rep., the Open Charge Alliance (OCA). https://www.openchargealliance.org (2020)
2. Connolly, D.T.: An improved annealing scheme for the qap. Eur. J. Oper. Res. **46**(1), 93–100 (1990)
3. Du, J., Leung, J.Y., Wong, C.S.: Minimizing the number of late jobs with release time constraint. J. Comb. Math. Comb. Comput. **11**(97), 107 (1992)
4. García-Álvarez, J., González, M.A., Vela, C.R.: Metaheuristics for solving a real-world electric vehicle charging scheduling problem. Appl. Soft Comput. **65**, 292–306 (2018)
5. Gong, L., Cao, W., Liu, K., Zhao, J.: Optimal charging strategy for electric vehicles in residential charging station under dynamic spike pricing policy. Sustain. Urban Areas **63**, 102474 (2020)
6. IEA: Global EV Outlook 2021: Accelerating ambitions despite the pandemic. International Energy Agency (IEA) (2021). www.iea.org
7. Kirkpatrick, S., Gelatt, C.D., Vecchi, M.P.: Optimization by simulated annealing. Science **220**(4598), 671–680 (1983)
8. Lipowski, A., Lipowska, D.: Roulette-wheel selection via stochastic acceptance. Phys. A **391**(6), 2193–2196 (2012)
9. Lundy, M., Mees, A.: Convergence of an annealing algorithm. Math. Program. **34**(1), 111–124 (1986)
10. Niu, L., Zhang, P., Wang, X.: Hierarchical power control strategy on small-scale electric vehicle fast charging station. J. Clean. Prod. **199**, 1043–1049 (2018)
11. Ogbu, F., Smith, D.K.: The application of the simulated annealing algorithm to the solution of the n/m/cmax flowshop problem. Comput. Operat. Res. **17**(3), 243–253 (1990)
12. Rahman, I., Vasant, P.M., Singh, B.S.M., Abdullah-Al-Wadud, M.: On the performance of accelerated particle swarm optimization for charging plug-in hybrid electric vehicles. Alex. Eng. J. **55**(1), 419–426 (2016)
13. Wang, Z., Jochem, P., Fichtner, W.: A scenario-based stochastic optimization model for charging scheduling of electric vehicles under uncertainties of vehicle availability and charging demand. J. Clean. Prod. **254**, 119886 (2020)
14. Wu, H., Pang, G.K.H., Choy, K.L., Lam, H.Y.: Dynamic resource allocation for parking lot electric vehicle recharging using heuristic fuzzy particle swarm optimization algorithm. Appl. Soft Comput. **71**, 538–552 (2018)
15. Yang, S.: Price-responsive early charging control based on data mining for electric vehicle online scheduling. Electric Power Syst. Res. **167**, 113–121 (2019)
16. Zhang, L., Li, Y.: Optimal management for parking-lot electric vehicle charging by two-stage approximate dynamic programming. IEEE Trans. Smart Grid **8**(4), 1722–1730 (2015)

Fine-Grained Cooperative Coevolution in a Single Population: Between Evolution and Swarm Intelligence

É. Lutton[1(✉)], S. Al-Maliki[2,3], J. Louchet[4], A. Tonda[1], and F. P. Vidal[2]

[1] UMR 518 MIA, INRAE, Palaiseau, France
{evelyne.lutton,alberto.tonda}@inrae.fr
[2] School of Computer Science and Electronic Engineering, Bangor University, Bangor, UK
{shatha.f.almaliki,f.vidal}@bangor.ac.uk
[3] Computer Science Department, College of Science, Basrh University, Basra, Iraq
[4] EFREI, Villejuif, France

Abstract. Particle Swarm Optimisation (PSO) and Evolutionary Algorithms (EAs) differ in various ways, in particular with respect to information sharing and diversity management, making their scopes of applications very diverse. Combining the advantages of both approaches is very attractive and has been successfully achieved through hybridisation. Another possible improvement, notably for addressing scalability issues, is cooperation. It has first been developed for co-evolution in EA techniques and it is now used in PSO. However, until now, attempts to make PSO cooperate have been based on multi-population schemes almost exclusively. The focus of this paper is set on single-population schemes, or fine-grained cooperation. By analogy with an evolutionary scheme that has long been proved effective, the fly algorithm (FA), we design and compare a cooperative PSO (coPSO), and a PSO-flavoured fly algorithm. Experiments run on a benchmark, the Lamp problem, show that fine-grained cooperation based on marginal fitness evaluations and steady-state schemes outperforms classical techniques when the dimension of the problem increases. These preliminary results highlight interesting future directions of research on fine-grained cooperation schemes, by combining features of PSO and FA.

1 Introduction

Swarm intelligence is a source of inspiration for many optimisation algorithms, for instance for PSO proposed by J. Kennedy and R. Eberhart in 1995 [22], Ant Colony Algorithms [16], Artificial Bee Colony Algorithms [20] or Bacterial Foragings [13]. The idea is to exploit the collective behaviour of a set of entities, the same way as natural populations (flocks of birds or ant colonies) search for food.

There is actually a proliferation of new techniques based on analogies to animal behaviour [38]. With respect to the ongoing debate about the originality

---

Supplementary Information The online version contains supplementary material available at https://doi.org/10.1007/978-3-031-42616-2_8.

and relevance of such a proliferation, we stress the fact that this contribution is not proposing yet another novel optimisation methodology, but making a point on two established heuristics that date back over 20 years and might look similar at first glance.

PSO is based on social interactions. The emerging collective behaviour results from a balance between following a leader and following an individual focus, thanks to inter-individual communications [31]. This mechanism is different from Evolutionary Algorithms (EAs) that rely on genetic transmission and natural selection analogies (birth, death and inheritance within a population). An important difference between them is how they manage diversity and share information[1], making them best fitted to different optimisation tasks [19].

Among other desirable features, scalability is a major concern. A way to deal with it is co-evolution, which was first developed for EA techniques [33] and starts to be experimented for PSO [6]. There are two major existing co-evolution schemes: mono- and multi-population [30], but as far as we know, only multi-population schemes are used in PSO [18,42].

This study investigates the differences and commonalities between intra-population communication in PSO and cooperative-co-evolution [12] as implemented in the FA [3,26,41]. This paper is organised as follows. After a rapid overview of the state of the art for PSO and cooperative PSO (Sect. 2), mono-population cooperative co-evolution and FA (Sect. 3), we propose a mono-population cooperative PSO (coPSO) and a new operator for the FA in Sect. 4. These schemes are compared on a cooperative-coevolution benchmark, the Lamp test case [39] in Sect. 5. The discussion and conclusions are given in Sect. 6.

2   From PSO to Cooperative PSO

Each entity of a PSO, called a particle, has a position in space and a velocity, that determines a random movement depending on the context. Velocities and positions are updated at each iteration using rules taking into account local and collective memories, mimicking respectively a cognitive and a social behaviour.

Similar to evolutionary techniques, the theoretical understanding of swarm intelligence is a formidable challenge: with very simple mechanisms, interactions of a large number of elements produce a nontrivial global dynamic. Besides experimental evidence that such a system is able to concentrate the population into optimal areas of a search space [35], theoretical results for convergence and convergence rates [31] exist and are based on simple PSO models. The parameter settings and the structure of the update rules clearly have a crucial influence on performance [37].

A canonical PSO can be described as follows [22]: each particle keeps track of its own best known position *pbest* and has also access at any time to the global swarm best known position *gbest*. An iteration loop is then implemented:

1. Particles are initialised with random positions and velocities.

---

[1] via inter-individual communications in PSO or genetic inheritance in EAs.

2. Best known positions are computed (according to the function to be optimised): $pbest_i$ for each particle i and $gbest$ for the whole swarm.
3. For each particle i, velocity v_i and position x_i are then updated (in vector notation, valid for any dimension of the search space):

$$v_i(t+1) = \omega v_i(t) + \varphi_p r_p(pbest_i - x_i) + \varphi_g r_g(gbest - x_i) \qquad (1)$$

$$x_i(t+1) = x_i(t) + v_i(t) \qquad (2)$$

where r_p and r_g are random values uniformly distributed between 0 and 1, ω is the inertia weight, φ_p and φ_g are the cognitive and social learning factors.
4. The process is repeated from Step 2 until a stopping criterion is met (e.g. stagnation, predefined level of fitness, or max. number of iterations).

The most common scheme, also called *gbest* **strategy**, corresponds to "fully informed" particles aware of the state of the whole population. In another important trend, called *lbest* **strategy**, each particle may only access local information [31]. The update rule is the same, except that in Eq. (1) *lbest*, a best *local* position, is used instead of *gbest*. This is more time-consuming as the neighbours of each particle (according to a given topology) have to be identified. This *lbest* scheme allows various subtleties to preserve diversity; neighbourhood topology has a strong influence on the performance of the Algorithms [23]. Topology may vary: the neighbourhood can be gradually enlarged according to a topological distance or a graph hierarchy, sometimes using adaptive strategies [31]. The *lbest* scheme is particularly useful in parallel implementations when communication between processors is limited [42]. However, it may cause trouble with high dimensional search spaces, as it relies on a distance measure which may become computationally expensive with large swarms, besides the fact that distance functions get less useful in high dimension spaces [27]. In this paper, we will focus on *gbest* strategies only.

Diversity is an important issue in PSO, to avoid premature convergence. For instance, dispersion and collision-avoiding mechanisms or repulsion mechanisms have been proposed [42]. It has to be noted that multi-population approaches have been developed for improving the management of diversity.

Cooperative PSO and multi-swarm models[2] have been developed for different purposes: to improve diversity [29,45], track multiple optima in multimodal or dynamic multimodal landscapes [11,31], address multi-objective problems [42], perform dynamic optimisation using adaptive strategies [8], handle constrained optimisation [36], or deal with large search spaces, by explicitly splitting the problem into interdependent sub-problems with smaller dimensions [42].

Bergh and Engelbrecht [6] were the first to use a cooperative scheme, in the style of Potter and De Jong [14,33] with separate sub-populations. Cooperation comes from the exchange of information between sub-populations, to build a composite fitness in the high-dimension problem. Usually the *gbest* particles of other sub-swarms are used to evaluate the particles of a sub-swarm. Fine tuning

---

[2] [18] defines cooperative search for any method as strategies that have several search modules running and exchanging information to improve search capability.

these algorithms is difficult [32], the choice of information to be exchanged and the synchronisation strategies deeply affect performance. It has been observed that "Increasing the number of cooperating swarms helps in improving the performance up to a certain limit, after which, the solution starts to deteriorate" [18].

Note that cooperative PSO developed until now corresponds to what we may call coarse-grained cooperation, *i.e.* the swarms or sub-swarms are explicitly separated: cooperation occurs at swarm-, not particle-level[3].

3   Fine Grained Cooperative Co-evolution

Co-evolution is an extension of standard EAs [30] that "distributes" the encoding of a solution onto several individuals. As a consequence the fitness of each individual depends on other individuals. An early example of this technique is the "Michigan approach" [44] for classifier systems, in which a single population of individuals, each being a rule, is evolved to collectively achieve a given task (rule-based machine learning). Another pioneering work is the multi-population approach of Potter and De Jong [33], later transferred to the PSO model. Co-evolution has actually been structured and exploited in optimisation in quite different ways, according the interacting behaviour, competitive versus cooperative [5,9,12,14,40,43] or the granularity of interaction: a single population of interbreeding individuals versus multiple interacting populations [30].

Various versions of **fine grained single-population cooperation** have been proposed: "Parisian Evolution" [12,17] in 2000 and more recently "Kaizen programming" [15], "FFX" [28] or "ϵ-lexicase survival" [24]. In [12], all individuals share the same representation, can exchange genetic material thanks to genetic operators and evolve together inside a single population. The EA loop then embeds an additional step at each generation for aggregating individuals to build a solution, evaluate it and distribute rewards to individuals. The idea is to exploit the evolution mechanism in a more parsimonious manner: where a traditional EA only keeps the best individual as an optimum solution at the end of the evolution (forgetting all precious information gathered by the population during its exploration of the search space), a Parisian approach tries to capitalise the full potential of an evolved population. It possess all usual features (e.g. mutation, crossover, and selection), but with two possible levels of fitness: **a local fitness** to assess the performance of a single individual (partial evaluation or local information) and **a global fitness** to assess the collective performance of the whole population. Maintaining diversity helps avoid degenerate solutions, e.g. when individuals gather in only a few areas of the search

---

[3] However, an application to the generation of improvised music [7] implements both types of cooperation, coarse and fine grained (this is not quite an optimisation, but rather an exploration task). It was performed with multi-swarms: each particle being a note (loudness, pulse and pitch of a MIDI event), each swarm a voice or instrument, and the whole system being considered as an improvising ensemble. Coherence is reached by self-organisation of particles and swarms.

space. Finally, a solution is built from a collation of individuals (sometimes with the concatenation of whole population). The way the fitness functions are constructed and the solution is extracted, are of course problem-dependent. Parisian Evolution has been successfully applied to various optimisation problems, such as text-mining [25], hand gesture recognition [21], complex interaction modelling in industrial agrifood processes [4,5], imaging problems such as computer stereo vision in robotics [26], tomography reconstruction in medical physics [2], and computer art [1].

A typical fine-grained cooperation is the Fly Algorithm (FA) [26]. First designed for stereovision applications, the Fly algorithm evolves a population of individuals called "flies". It uses an "inverse problem" approach where conventional approaches to stereovision use primitive extraction, pattern matching and calculation of disparities [26]. In the original version, a fly is defined as a 3-D point (x, y, z). A population of flies is initialised in the field of view common to at least two cameras, then evolved using a classical Evolutionary Strategy, guided by the flies' fitness values. The solution is given by the whole population (or a subset of the population), concentrated on the visible surfaces of the objects in the scene [10]. The fitness of a fly is a measurement of the consistency of its projections on the cameras. Classical operators – mutation, optional CMX crossover, immigration (introducing brand new flies) and tournament selection – are most commonly used.

4   Fine-Grained Optimisation Based on PSO and FA

Particle Swarm Optimisation versus Fly Algorithm
Besides the narrative attached to each scheme (communications and social behaviour versus genealogical features transmission and selection mechanisms), PSO and FA share obvious features, and a parallel can be drawn between flies mutations and particle movements, but this actually leads to a different balance between diversification and intensification [19]. In particular, selection is not used in PSO, although it is an explicit intensification mechanism. Additionally, diversity preservation mechanisms are more explicit and tunable in FA, with the help of an "immigration" operator that introduces a proportion of purely random flies in the current population. We propose hereafter two different lines for mutual cross-fertilisation (i) implementing the PSO algorithm using the Parisian approach, and (ii) introducing the same information sharing mechanism as in PSO into the FA.

A Cooperative PSO: coPSO
A coPSO, in terms of fine grained approach, consists in evolving, within a single swarm, particles that carry only a small part of a solution. At each iteration of the algorithm it is necessary to aggregate the particles of the swarm (or a selected part of it) to build the problem solution. As for FA, there are now two levels of objective functions, an optional global one computed on the whole swarm and a local one computed for each particle. The local fitness function is used to update *pbest*. Due to the distributed nature of the approach, the social

learning factor (φ_g in Eq. 1) is set to 0 as it makes no sense to follow the global best particle (*gbest*). Equation 2 remains the same. In the experiments below, a marginal fitness[4] is used at the local level.

FA as a Swarm: SFA

To introduce a "PSO-like" information sharing mechanism within a FA, we built an additional operator, the *genealogical mutation*. The idea is, for each individual, to keep track of the best of its ancestors, according to the genealogy due to the genetic operators. Additionally, an extra vector similar to the velocity in PSO is attached to each fly. When a genealogical mutation is triggered, the velocity and position of offspring are updated using Eqs. 1 and 2. For the same reason as above, φ_g is set to zero. Note that this operator tends to focus the search of a fly into the direction of its *pbest*. However, it may be too restrictive (i) at the start of the optimisation when no knowledge is available, and (ii) at the end of optimisation when the result needs to be refined. This is why an adaptive mutation scheme has been built.

Adaptive Mutation

The adaptive *genetic bi-operator*, concurrently assesses two different genetic operators (here Gaussian mutation and genealogical mutation) so that the most successful operator in generating good offspring is favoured. Both operators are initially given an equal probability of occurrence. Their success rates are checked at regular intervals to adjust their probabilities. The update rule is multiplicative as for the famous $1/5^{\text{th}}$ rule [34].

Each operator has i) a counter to keep track of how many times it has been applied and ii) an accumulator that keeps track of how many times it has been successful. This accumulator is incremented if the marginal fitness of the newly created fly is positive, decremented if negative. The success rate of an operator is its accumulator divided by its counter. The probability of the most successful operator over the last period is increased at the expense of the other one. The probabilities are then clamped in the range 10%-90% to make sure that the least successful operator retains a chance to be picked up.

5   Experimental Analysis on a Toy Problem

A Toy Problem for Cooperative-Coevolution: The Lamps

There are few benchmarks designed for cooperative co-evolutionary algorithms. *The Lamps* [39] is one of the toy problems available: the basic premise is to optimally place a set of circles (*lamps*) of given radius, so that they completely cover a square field. The fitness function rewards each lamp separately, and also provides a global reward that depends on the overall placement of all lamps. While each single lamp can be optimally placed on the square field, so that it

---

[4] Positive or negative contribution of the individual to the global fitness, *i.e.* the difference between the fitness of the population, when complete or deprived from this particular individual. This concept has been successfully used in various applications, see for instance [2]. In the absence of additional information at the local level for building a specific "local fitness", marginal fitness is a convenient option.

lits as much area as possible, it is interesting to notice that sometimes individual lamps with sub-optimal positions (e.g. part of their area falls outside of the field) can significantly improve the global reward (see Fig. 1). This simple toy problem only has one parameter, the ratio between the radius of a circle/lamp and the side of the square field (*i.e.* the ratio between the surface of a lamp and the surface of the field), $problem\_size = \frac{area\_room}{area\_lamp}$. With higher parameter values, more lamps are needed with more placement possibilities, making the benchmark more challenging. A further difficulty can be added by introducing penalties for overlapping lamps.

$$fitness = \frac{area\_enlightened}{total\_area} - W.\frac{area\_overlap}{total\_area} \qquad (3)$$

The fitness of a candidate involves the total area enlightened and the number of lamps used. A weight W sets the balance with the overlapping term, see Eq. (3). Best solutions maximise the illuminated area whilst minimising the number of lamps to cover the whole area. Tonda *et al.* showed that traditional approaches based on genetic operators are competitive when the search space is relatively small, *i.e.*, for Lamps problem size less than 10 [39]. For more complex problems, the Parisian approach outperformed the other algorithms tested.

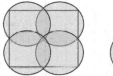

Fig. 1. Arrangement of a set of four lamps to enlighten a square field. (**left**) The lamps completely cover the square field, but part of their own area is outside of the square itself. (**right**) One of the lamps is now completely inside the square, but the global solution is unable to completely cover the square.

Experimental Setup

The Lamps problem with increasing sizes (3, 5, 10, 20, 100, and 500) has been used for benchmarking the scalability of six algorithms:

- A traditional PSO with no algorithmic enhancement, as a baseline for comparison (labelled PSO in the tables and figures below);
- The coPSO algorithm (labelled coPSO in the tables and figures);
- A steady state FA with marginal fitness, threshold selection, varying population size using mitosis and slaughtering/culling, 30% of immigration and 70% of Gaussian mutation (labelled FA);
- A steady state FA as above but with 30% of immigration, 35% of Gaussian mutation, and 35% of Genealogical mutation (labelled SFA35);
- As above but with 30% of immigration, and 70% of Genealogical mutation (labelled SFA70);

– As above but with 30% of immigration, and 70% of genetic bi-operator with both Gaussian and Genealogical mutation (labelled SFA-bi operator).

The lamp radius is 8 and $W = 1$ (Eq. 3) to match the value initially used in [39]. Algorithms 1 and 2 show the skeleton of FA and coPSO implementations, displayed side-by-side to highlight similarities and differences. The structure of

```
// Read problem specific data
                // Set the algorithm
Initialisation
// Create the initial population of n
   individuals
repeat n times
   Create a fly at a random position in
      the search space;

   Add the fly to the population;
   Add the fly's contribution to the
      population's fitness;
end
Compute the global fitness;

repeat                  // Optimisation loop
   repeat n times
      repeat             // Select a bad fly
         i ← Random(0, n − 1);
         MF(i) ← Marginal fitness of
            Fly i;
      until MF(i) ≤ 0;
      Remove Fly(i)'s contribution
         from the population's;

      Compute the global fitness;

      Select genetic operator;
      if Genetic operator is
         immigration then
         │  Replace Fly(i) with a random
         │     fly in the search space;
      else          // Mutation is used
         │  repeat // Select a good fly
         │     j ← Random(0, n − 1);
         │     MF(j) ← Marginal
         │        fitness of Fly j;
         │  until MF(j) > 0;
         │  Copy Fly(j)'s genes into
         │     Fly(i)'s;
         │  Randomly mutate Fly(i)'s
         │     genes;
      end
      Add Fly(i)'s contribution to
         global fitness;

      Compute the global fitness;
   end
until Convergence;
Iteratively eliminate bad flies;
Convert the population of flies into
   problem specific answer;
```

Algorithm 1: Steady state FA

```
// Read problem specific data
                // Set the algorithm
Initialisation
// Create the initial swarm of n
   particle
repeat n times
   Create a particle at a random position
      in the search space;
   Initialise the particle's velocity;
   Add the particle to the swarm;
   Add the particle's contribution to the
      swarm's;
end
Compute the global fitness;

repeat                  // Optimisation loop
   foreach Particle p_i ∈ Swarm do




      Remove p_i's contribution from
         the swarm's;


















      Update the p_i's velocity;
      Update the p_i's position;
      Compute the global fitness;
      Compute p_i's local fitness
         (Marginal fitness)
      Update p_i's lbest if needed
   end
until Convergence;
Iteratively eliminate bad particles;
Convert the swarm of particles into
   problem specific answer;
```

Algorithm 2: Cooperative PSO.

the algorithms is fairly similar but coPSO lacks natural selection for killing and breeding. The mutation in FA and the position update in coPSO are similar in the sense that they both move an individual or particle from its current position. Algorithmic enhancements such as varying population or swarm size are not shown to improve the readability of the pseudocode. In our experiment, we added an extra loop so that each time stagnation is detected slaughtering/culling and mitosis are alternatively triggered. In the slaughtering/culling step, bad flies or particles are eliminated so that there are only good flies or particles left. If triggering slaughtering/culling and mitosis does not help the population or swarm improve the global fitness over N iterations (stagnation), the optimisation ends and the problem solution is extracted. Our main stopping criterion is thus stagnation.

N is set to 5 for coPSO, FA and SFA. However, it was empirically determined that this number was far too low for PSO, which is why we use 50 in our experiments. An additional stopping criterion is the maximum number of iterations in case an algorithm fails to converge towards a solution. All parameters are provided in Table 1.

Each experiment is repeated 100 times using a supercomputer to gather statistically meaningful results[5]. That is 100 runs × 6 problem sizes × 6 algorithms = 3600 optimisation processes in total. Each algorithm records the global fitness of the solution it provided, and how many lamps needed to be created and tested before a solution was accepted. This number is linearly proportional to the computational power that was required to find the solution.

Results and Discussion[6]

Quantitative results are given in Table 2. It highlights for each problem size which algorithm(s) provides solutions significantly better ($p < 0.05$) than the other algorithms. From the table it is clear that PSO performs best with small problem sizes but collapses rapidly. It is also computationally intensive compared to FAs. Figures 2 and 3 are a visualisation of these data in terms of global fitness and computing time versus problem size in log scale.[7]

With small problem sizes, FA does not perform quite as well as PSO; SFA is comparable to FA though a little less performing, and coPSO does not perform well at all. With larger problem sizes, Fig. 2 shows that PSO collapses; FA stabilises, taking advantage of its scalability; coPSO starts a shy improvement, showing that communication is only beneficial with a larger problem size. Figure 3 clearly shows that PSO and coPSO are not as efficient as FA and SFA. On both figures FA and the 3 variants of SFA are hard to distinguish when the problem size increases. A zoomed scatterplot (Fig. 4) of global fitness versus computational effort (number of lamps created) gives a more precise comparison for FA and SFA. FA's performance decreases when the problem size increases to become quite close to SFA35's and SFA70's. SFA-bi operator is, however, more consistent and starts to slightly outperform FA in terms of computing requirements ($p < 0.05$).

---

[5] Except for the largest instance (size 500) for which only 50 runs were done.

[6] **Reproducibility:** code available at http://doi.org/10.5281/zenodo.7101160.

[7] A synthetic scatterplot is also provided in https://evelyne-lutton.fr/Lutton_EA2022-Additional.pdf for assessing the balance between both measurements.

Table 1. Summary of the algorithms' parameters.

	PSO	FA	coPSO	SFA35	SFA70	SFA-bi operator
Initial number of particles/individuals:	$\sqrt{\frac{\overline{FA}}{3\times\text{pb size}}}$	3 × pb size	3 × pb size	3 × pb size	3 × pb size	3 × pb size
Lamps per particle/individual:	3 × pb size	1	1	1	1	1
W in Eq. 3:	1	1	1	1	1	1
Immigration probability (%):	N/A	30	N/A	30	30	30
Gaussian mutation probability (%):	N/A	70	N/A	35	0	varying
Genealogical mutation probability (%):	N/A	0	N/A	35	70	varying
Initial Gaussian mutation factor (pixels):	N/A	16	N/A	16	N/A	16
Decrease of mutation factor per generation:	N/A	0.016 pixel	N/A	0.016 pixel	N/A	0.016 pixel
ω in Eq. 1:	$\frac{1}{2\times\log(2)}$	N/A	$\frac{1}{2\times\log(2)}$	$\frac{1}{2\times\log(2)}$	$\frac{1}{2\times\log(2)}$	$\frac{1}{2\times\log(2)}$
φ_p in Eq. 1:	$\frac{1}{2}+\log(2)$	N/A	$\frac{1}{2}+\log(2)$	$\frac{1}{2}+\log(2)$	$\frac{1}{2}+\log(2)$	$\frac{1}{2}+\log(2)$
φ_g in Eq. 1:	$\frac{1}{2}+\log(2)$	N/A	0	0	0	0
Stopping criteria						
1) No improvement over the last :	50 iterations	5 iterations	5 iterations	5 iterations	5 iterations	5 iterations
2) Max # of gen. or iterations:	500	500	500	500	500	500

For each problem size, $\overline{FA}$ is the average number of lamps created over 100 runs of FA to reach the problem solution

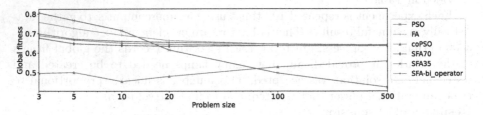

Fig. 2. Comparison in terms of global fitness (maximisation).

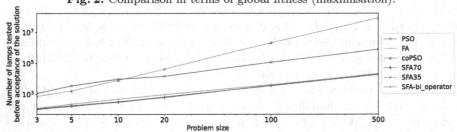

Fig. 3. Comparison in terms of computational requirement. This is a value that should be as small as possible.

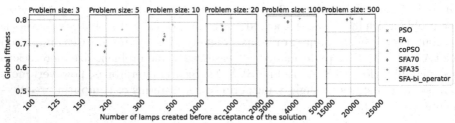

Fig. 4. Performance comparison in terms of effectiveness (highest global fitness) and efficiency (smallest number of tested lamps) zoomed onto FA and SFA. The best algorithms are in the top-left corner of the plots.

Table 2. Results of the experiments. Values for algorithms marked in **bold** are significantly better ($p < 0.05$) than the others for the same problem size. Values in *italics* highlight cases where the best performance is equally achieved by two or more algorithms (non separable, with $p > 0.05$).

Problem size	Evolution	Global fitness	Lamps created before acceptance
3	**PSO**	**80.58% ± 6.84**	1.18e+03 ± 473.47
3	FA	75.81% ± 4.22	1.32e+02 ± 100.75
3	coPSO	49.66% ± 11.58	7.63e+02 ± 1560.97
3	SFA70	67.74% ± 7.59	1.23e+02 ± 100.48
3	SFA35	69.03% ± 7.56	1.08e+02 ± 79.37
3	SFA-bi operator	69.87% ± 7.71	1.18e+02 ± 83.20
5	**PSO**	**76.74% ± 4.98**	3.63e+03 ± 1296.64
5	FA	72.71% ± 3.51	2.49e+02 ± 138.45
5	coPSO	50.09% ± 8.51	1.73e+03 ± 1274.77
5	SFA70	64.82% ± 5.23	1.95e+02 ± 152.83
5	SFA35	66.80% ± 4.83	1.99e+02 ± 142.94
5	SFA-bi operator	67.21% ± 4.74	1.75e+02 ± 129.27
10	**PSO**	**71.44% ± 4.48**	9.89e+03 ± 2843.23
10	FA	69.11% ± 3.28	5.15e+02 ± 301.90
10	coPSO	45.51% ± 7.39	8.39e+03 ± 6427.39
10	SFA70	63.91% ± 4.02	3.31e+02 ± 208.52
10	SFA35	65.13% ± 3.90	3.55e+02 ± 225.33
10	SFA-bi operator	66.01% ± 3.22	3.49e+02 ± 205.07
20	PSO	61.52% ± 4.46	1.50e+04 ± 3893.98
20	**FA**	**66.39% ± 2.56**	9.97e+02 ± 536.03
20	coPSO	48.12% ± 6.07	4.37e+04 ± 27148.87
20	SFA70	63.49% ± 2.81	6.77e+02 ± 290.76
20	SFA35	64.46% ± 2.69	6.51e+02 ± 322.55
20	SFA-bi operator	65.47% ± 2.26	7.09e+02 ± 253.34
100	PSO	49.74% ± 2.26	1.20e+05 ± 34739.96
100	*FA*	*64.47% ± 1.11*	4.33e+03 ± 1664.40
100	coPSO	52.86% ± 4.53	2.14e+06 ± 1215718.00
100	SFA70	63.78% ± 1.29	3.85e+03 ± 854.81

(*continued*)

Table 2. (*continued*)

Problem size	Evolution	Global fitness	Lamps created before acceptance
100	SFA35	63.81% ± 1.42	3.86e+03 ± 912.96
100	*SFA-bi operator*	*64.57% ± 1.18*	3.71e+03 ± 842.15
500	PSO	42.43% ± 1.37	7.94e+05 ± 252250.84
500	*FA*	*63.73% ± 0.56*	2.22e+04 ± 7000.81
500	coPSO	56.04% ± 3.09	9.00e+07 ± 40418907.90
500	*SFA70*	*63.66% ± 0.61*	2.02e+04 ± 3852.50
500	*SFA35*	*63.81% ± 0.57*	2.07e+04 ± 4109.04
500	*SFA-bi operator*	*64.03% ± 0.54*	2.01e+04 ± 3935.58

6   Conclusions

The previous experiments are a first try with fine-grained cooperative swarms. For the moment this has been reached with a single swarm in which social communications have been cut (the *gbest* position has no influence on local rules). Together with a convenient formulation of the problem (which information is carried by a particle), this rough strategy (coPSO) is able to drive the full swarm into a good solution, while having better scalability than the classical *gbest* PSO. Marginal fitness is actually an indirect way to implement some social communication, as it evaluates the contribution of a particle with respect to the whole swarm. Less efficient than FA, a mature technique of fine-grained cooperative based on EA, this simple coPSO however exhibits interesting scalability properties (positive slope on Fig. 2 for large problem sizes).

Future improvements of this strategy can follow different lines. A first one could be distance-based *lbest* strategies, but possibly limited in high dimensions. Another line, sketched in this paper, is a combination of features from Evolutionary Algorithms (life and death, genetic transmission) and swarms (internal memory and social communication in the group). SFA is an attempt to add a memory to the flies, in the same spirit as coPSO, as an inter-generational transmission of information. The experiments displayed above prove that these inter-generational communications improve the scalability of FA. Making the balance of the SFA mutations adaptive also yields important information about the efficiency of these operators during the runs (see also supplementary material).

Future work on this topic will aim at exploring the combinations of PSO and FA, and extending the experiments to other benchmarks and real problems.

Acknowledgements. We would like to thank Supercomputing Wales for the supercomputer used to generate all the experimental results (https://www.supercomputing.wales/).

References

1. Abbood, Z.A., Vidal, F.P.: Fly4Arts: evolutionary digital art with the Fly algorithm. ISTE Arts Sci. **17**–1(1), 11–16 (2017). https://doi.org/10.21494/ISTE.OP.2017.0177
2. Ali Abbood, Z., Lavauzelle, J., Lutton, E., Rocchisani, J.M., Louchet, J., Vidal, F.P.: Voxelisation in the 3-D fly algorithm for PET. Swarm Evol. Comput. **36**, 91–105 (2017). https://doi.org/10.1016/j.swevo.2017.04.001
3. Ali Abbood, Z., Vidal, F.P.: Basic, dual, adaptive, and directed mutation operators in the fly algorithm. In: Lutton, E., Legrand, P., Parrend, P., Monmarché, N., Schoenauer, M. (eds.) EA 2017. LNCS, vol. 10764, pp. 100–114. Springer, Cham (2018). https://doi.org/10.1007/978-3-319-78133-4_8
4. Barrière, O., Lutton, E.: Experimental Analysis of a Variable Size Mono-population Cooperative-Coevolution Strategy, pp. 139–152. Springer (2009). https://doi.org/10.1007/978-3-642-03211-0_12
5. Barrière, O., Lutton, E., Wuillemin, P.: Bayesian network structure learning using cooperative coevolution. In: GECCO, pp. 755–762 (2009). https://doi.org/10.1145/1569901.1570006
6. Bergh, F., Engelbrecht, A.: A cooperative approach to particle swarm optimization. IEEE Trans. Evolut. Comput. **8**, 225–239 (2004). https://doi.org/10.1109/TEVC.2004.826069
7. Blackwell, T.: Swarm music: improvised music with multi-swarms. In: Symposium on Artificial Intelligence and Creativity in Arts and Science, pp. 41–49 (2003)
8. Blackwell, T., Branke, J.: Multi-swarm optimization in dynamic environments. In: Raidl, G.R., et al. (eds.) EvoWorkshops 2004. LNCS, vol. 3005, pp. 489–500. Springer, Heidelberg (2004). https://doi.org/10.1007/978-3-540-24653-4_50
9. Bongard, J., Lipson, H.: Active coevolutionary learning of deterministic finite automata. J. Mach. Learn. Res. **6**, 1651–1678 (2005)
10. Boumaza, A.M., Louchet, J.: Mobile Robot Sensor Fusion Using Flies, pp. 357–367 (2003). https://doi.org/10.1007/3-540-36605-9_33
11. Brits, R., Engelbrecht, A., van den Bergh, F.: Scalability of niche PSO. In: Proceedings of the IEEE Swarm Intelligence Symposium, Indianapolis, Indiana, USA, 24–26 April, pp. 228–234 (2003)
12. Collet, P., Lutton, E., Raynal, F., Schoenauer, M.: Polar IFS + Parisian genetic programming = efficient IFS inverse problem solving. Genetic Program. Evolvable Mach. J. **1**(4), 339–361 (2000)
13. Das, S., Biswas, A., Dasgupta, S., Abraham, A.: Bacterial Foraging Optimization Algorithm: Theoretical Foundations, Analysis, and Applications, pp. 23–55 (2009). https://doi.org/10.1007/978-3-642-01085-9_2
14. De Jong, E.D., Stanley, K.O., Wiegand, R.P.: Introductory tutorial on coevolution. In: GECCO 2007, London, UK (2007)
15. De Melo, V.V.: Kaizen programming. In: GECCO 2014: Proceedings of the 2014 Conference on Genetic and Evolutionary Computation, 12–16 Jul, pp. 895–902. ACM, Vancouver, BC, Canada (2014). https://doi.org/10.1145/2576768.2598264
16. Dorigo, M., Birattari, M., Stutzle, T.: Ant colony optimization. IEEE Comput. Intell. Mag. **1**(4), 28–39 (2006)

17. Dunn, E., Olague, G., Lutton, E.: Parisian camera placement for vision metrology. Pattern Recog. Lett. **27**(11), 1209–1219 (2006). https://doi.org/10. 1016/j.patrec.2005.07.019,https://www.sciencedirect.com/science/article/pii/ S016786550500334X, evolutionary Computer Vision and Image Understanding

18. El-Abd, M., Kamel, M.S.: A taxonomy of cooperative particle swarm optimizers. Int. J. Comput. Intell. Res. **4** (2008). https://doi.org/10.5019/j.ijcir.2008.133

19. Kachitvichyanukul, V.: Comparison of three evolutionary algorithms: Ga, pso, and de. Indus. Eng. Manag. Syst. **12**, 215–223 (2012). https://doi.org/10.7232/iems. 2012.11.3.215

20. Karaboga, D., Gorkemli, B., Ozturk, C., Karaboga, N.: A comprehensive survey: artificial bee colony (ABC) algorithm and applications. Artif. Intell. Rev. **42**, 21–57 (2014). https://doi.org/10.1007/s10462-012-9328-0

21. Kaufmann, B., Louchet, J., Lutton, E.: Hand posture recognition using real-time artificial evolution. In: Di Chio, C., et al. (eds.) EvoApplications 2010. LNCS, vol. 6024, pp. 251–260. Springer, Heidelberg (2010). https://doi.org/10.1007/978-3-642-12239-2_26

22. Kennedy, J., Eberhart, R.: Particle swarm optimization. In: Proceedings of ICNN 1995 - International Conference on Neural Networks, vol. 4, pp. 1942–1948 (Nov 1995). https://doi.org/10.1109/ICNN.1995.488968

23. Kennedy J, M.R.: Population structure and particle swarm performance. In: CEC, Honolulu, HI, USA, 22–25 Sept, pp. 1671–1676 (2002)

24. La Cava, W., Moore, J.: A general feature engineering wrapper for machine learning using ε-Lexicase survival. In: McDermott, J., Castelli, M., Sekanina, L., Haasdijk, E., García-Sánchez, P. (eds.) EuroGP 2017. LNCS, vol. 10196, pp. 80–95. Springer, Cham (2017). https://doi.org/10.1007/978-3-319-55696-3_6

25. Landrin-Schweitzer, Y., Collet, P., Lutton, E.: Introducing lateral thinking in search engines. Genet. Program Evolvable Mach. **7**(1), 9–31 (2006). https://doi. org/10.1007/s10710-006-7008-z

26. Louchet, J.: Using an individual evolution strategy for stereovision. Genet. Program Evolvable Mach. **2**(2), 101–109 (2001). https://doi.org/10.1023/A: 1011544128842

27. Marimont, R.B., Shapiro, M.B.: Nearest neighbour searches and the curse of dimensionality. IMA J. Appli. Mathem. **24**(1), 59–70 (1979). https://doi.org/10.1093/ imamat/24.1.59

28. McConaghy, T.: FFX: Fast, scalable, deterministic symbolic regression technology. In: Riolo, R., Vladislavleva, E., Moore, J.H. (eds.) Genetic Programming Theory and Practice IX, chap. 13, pp. 235–260. Genetic and Evolutionary Computation. Springer, Ann Arbor, USA (12–14 May 2011). http://trent.st/content/2011-GPTP-FFX-paper.pdfhttps://doi.org/10.1007/978-1-4614-1770-5_13

29. Niu, B., Zhu, Y., He, X.: Multi-population cooperative particle swarm optimization. In: Capcarrère, M.S., Freitas, A.A., Bentley, P.J., Johnson, C.G., Timmis, J. (eds.) ECAL 2005. LNCS (LNAI), vol. 3630, pp. 874–883. Springer, Heidelberg (2005). https://doi.org/10.1007/11553090_88

30. Ochoa, G., Lutton, E., Burke, E.K.: Cooperative royal road functions. In: Evolution Artificielle, Tours, France, 29–31 October (2007)

31. Poli, R., Kennedy, J., Blackwell, T.: Particle swarm optimization. Swarm Intell. **1**(1), 33–57 (2007). https://doi.org/10.1007/s11721-007-0002-0

32. Popovici, E., Bucci, A., Wiegand, R.P., De Jong, E.D.: Coevolutionary Principles, pp. 987–1033. Springer (2012). https://doi.org/10.1007/978-3-540-92910-9_31

33. Potter, M.A., De Jong, K.A.: A cooperative coevolutionary approach to function optimization. In: Davidor, Y., Schwefel, H.-P., Männer, R. (eds.) PPSN 1994. LNCS, vol. 866, pp. 249–257. Springer, Heidelberg (1994). https://doi.org/10.1007/3-540-58484-6_269

34. Schwefel, H.P.: Evolution and Optimum Seeking: The Sixth Generation. John Wiley & Sons Inc., USA (1993)

35. Shi, Y., Eberhart, R.C.: Empirical study of particle swarm optimization. In: Congress on Evolutionary Computation, CEC 1999, vol. 3, pp. 1945–1950 (July 1999). https://doi.org/10.1109/CEC.1999.785511

36. Shi, Y., Krohling, R.A.: Co-evolutionary particle swarm optimization to solve min-max problems. In: CEC, vol. 2, pp. 1682–1687 (2002)

37. Shi, Y., Eberhart, R.C.: Parameter selection in particle swarm optimization. In: Porto, V.W., Saravanan, N., Waagen, D., Eiben, A.E. (eds.) EP 1998. LNCS, vol. 1447, pp. 591–600. Springer, Heidelberg (1998). https://doi.org/10.1007/BFb0040810

38. Sörensen, K.: Metaheuristics-the metaphor exposed. Int. Trans. Oper. Res. $22(1)$, 3–18 (2015)

39. Tonda, A., Lutton, E., Squillero, G.: Lamps: A test problem for cooperative coevolution. In: NICSO, vol. 387, pp. 101–120. Springer (2011). https://doi.org/10.1007/978-3-642-24094-2_7

40. Vidal, F.P., Lazaro-Ponthus, D., Legoupil, S., Louchet, J., Lutton, E., Rocchisani, J.M.: Pet reconstruction using a cooperative coevolution strategy. In: Proceedings of the IEEE Medical Imaging Conference 2009. IEEE, Orlando, Florida (Oct 2009)

41. Vidal, F.P., Lutton, E., Louchet, J., Rocchisani, J.-M.: Threshold selection, mitosis and dual mutation in cooperative co-evolution: application to medical 3D tomography. In: Schaefer, R., Cotta, C., Kołodziej, J., Rudolph, G. (eds.) PPSN 2010. LNCS, vol. 6238, pp. 414–423. Springer, Heidelberg (2010). https://doi.org/10.1007/978-3-642-15844-5_42

42. Wang, D., Tan, D., Liu, L.: Particle swarm optimization algorithm: an overview. Soft Computing (Jan 2017). https://doi.org/10.1007/s00500-016-2474-6

43. Wiegand, R.P., Potter, M.A.: Robustness in cooperative coevolution. In: Proceedings of GECCO, Seattle, Washington, USA, pp. 369–376 (2006). https://doi.org/10.1145/1143997.1144063

44. Wilson, S.T., Goldberg, D.E.: A Critical Review of Classifier Systems. In: Third International Conference on Genetic Algorithms, pp. 244–255 (1989)

45. Zhang, H.: A Newly Cooperative PSO - Multiple Particle Swarm Optimizers with Diversive Curiosity, MPSOα/DC, pp. 69–82. Springer, Netherlands (2011). https://doi.org/10.1007/978-94-007-0286-8_7

One-Class Ant-Miner: Selection of Majority Class Rules for Binary Rule-Based Classification

Naser Ghannad[1,2]([✉]) [iD], Roland de Guio[1,2] [iD], and Pierre Parrend[1,3] [iD]

[1] ICube (Laboratoire des sciences de l'ingénieur, de l'informatique et de l'imagerie), UMR 7357, Université de Strasbourg, CNRS, 67000 Strasbourg, France
[2] INSA de Strasbourg, Strasbourg, France
{Naser.Ghannad,Roland.Deguio}@insa-strasbourg.fr
[3] EPITA, 5, Rue Gustave Adolphe Hirn, Strasbourg, France
Pierre.Parrend@epita.fr

Abstract. In recent years, high-performance models have been introduced based on deep learning; however, these models do not have high interpretability to complement their high efficiency. Rule-based classifiers can be used to obtain explainable artificial intelligence. Rule-based classifiers use a labeled dataset to extract rules that express the relationships between inputs and expected outputs. Although many evolutionary and non-evolutionary algorithms have developed to solve this problem, we hypothesize that rule-based evolutionary algorithms such as the AntMiner family can provide good approximate solutions to problems that cannot be addressed efficiently using other techniques. This study proposes a novel supervised rule-based classifier for binary classification tasks and evaluates the extent to which algorithms in the AntMiner family can address this problem. First, we describe different versions of AntMiner. We then introduce the one-class AntMiner (OCAntMiner) algorithm, which can work with different imbalance ratios. Next, we evaluate these algorithms using specific synthetic datasets based on the AUPRC, AUROC, and MCC evaluation metrics and rank them based on these metrics. The results demonstrate that the OCAntMiner algorithm performs better than other versions of AntMiner in terms of the specified metrics.

Keywords: AntMiner · Evolutionary algorithm · Rule-based classifier · Ant colony classification · Imbalanced dataset · Binary classification · Synthetic datasets

1 Introduction

In recent years, various highly scalable models have been developed using deep learning [8,10] and other machine learning models such as XGBoost [5]. Most of these models are black-box models that are not interpretable by users. A few of these models specify the importance of features to help users understand which features are more helpful for predicting classes [15]. However, they do not provide explicit relationships that allow human users to understand the

P. Legrand et al. (Eds.): EA 2022, LNCS 14091, pp. 118–132, 2023.
https://doi.org/10.1007/978-3-031-42616-2_9

relationships between input and output variables, unlike a white-box model. Because rule-based classifiers [1] explicitly rely on individual variables in the original data, they are powerful candidates for constructing white-box models. Rule-based classifiers extract rules to explain the effects of individual variables on a given class, as follows:

$$IF\ (Conditions)\ THEN\ (Consequent) \tag{1}$$

Conditions are combinations of propositions for different input variables (terms) bound by a logical conjunction (AND). The result of these combinations is the consequent (i.e., classes). In this study, we focused on ordered-rule-based classifiers. There are two methods for extracting rules: direct and indirect. In direct methods, the algorithm directly operates on the data and extracts rules from the data, as seen in the RIPPER [7], CN2 [6], PART [14], and RISE [11] algorithms, or uses evolutionary algorithms such as the ant colony algorithm (AntMiner) [27] to extract rules. In indirect methods, a classifier is first applied to the data and then another method extracts rules from the classifier. Such methods include the C4.5, J48 [30,31], random tree [29], and REPTree [33] algorithms. These tree-based algorithms first use a decision tree to classify data. Rules are then extracted from the trees provided by the algorithms.

The goal of this study was to develop an algorithm to extract rules from provided datasets that works well with both imbalanced (i.e., when the amount of data in one class is significantly less than the amount of data in another class) and balanced datasets, and also determine a suitable method for the ranking algorithms based on datasets with various imbalance ratios. Additionally, considering the lack of suitable datasets for evaluating rule-based algorithms, we aimed to generate datasets containing all possible instances for generating output classes so that we could determine which algorithms could achieve the highest values for the defined metrics with the availability of all possible instances and absence of noise. Also, with these data, we can find out which algorithm is over-fitted or under-fitted on the data. The remainder of this article is organized as follows. Section 2 provides an overview of evolutionary approaches. Section 3 defines the developed one-class AntMiner (OCAntMiner) algorithm, and Sect. 4 provides an evaluation of the considered algorithms. Section 5 presents the evaluation results. Finally, Sect. 6 concludes this article.

2 Related Work

Evolutionary algorithms (EAs) are used in optimization problems. They represent a subset of evolutionary computations [34]. The EA approach is inspired by biological evolution and uses operations such as mutation, recombination (crossover), and selection. A population evolves with the goal of maximizing a given evaluation function. The main EAs used for learning rules are the learning classifier system (LCS) [4] and AntMiner algorithms [27]. The LCS was introduced by John Holland [17], and genetic algorithms were used to extract rules. In contrast, AntMiner uses a simulated ant colony as a probabilistic approach to solve graph problems and find the best path to optimize an evaluation function. Ants choose their paths based on the amount of pheromones along paths, which represents how many times a path has been selected successfully before. Releasing pheromones in an environment (along paths) is a method for ants to communicate. The path with the most pheromones should be shorter than the others according to the current state of the search [12].

2.1   Evolution of AntMiner

Several versions of AntMiner have been released. We focus on the most important releases. As mentioned previously, the first version of AntMiner was introduced by Parpinelli [27,28]. The second version was AntMiner2 [18], which introduced a novel heuristic function. AntMiner+ [19] defined a directed acyclic graph as a new environment for ants to select their paths more efficiently when the AntMiner environment is fully connected. This means that AntMiner ants must choose from all nodes at each decision point, whereas AntMiner+ ants must choose only from nodes corresponding to a single variable. The continuous AntMiner (cAntMiner) [21,22] follows the concepts of AntMiner+ [19] and removes the discretization step by dynamically determining the cutoff values for continuous variables by selecting corresponding entropy minimization values. Ant-Tree-Miner [26] uses the ant colony optimization algorithm to learn decision trees. The Ant-Tree-Miner algorithm follows the traditional divide-and-conquer approach. It uses a stochastic process based on heuristic information and pheromone values during tree construction to define the nodes (attributes) of trees instead of applying greedy deterministic selection. In the Pittsburgh cAnt-Miner (cAntMinerPB) [20,23], a search for the best list of rules is performed, whereas in AntMiner, a search for the best single rule is performed in each step of the sequential coverage process. In other words, in cAntMinerPB, the search is guided by the quality of a candidate list of rules, whereas in AntMiner, it is guided by the quality of a single rule. In the unordered cAntMinerPB (UCAnt-MinerPB) [24,25], the underlying concept is the same as that of cAntMinerPB, but the goal is to generate an unordered set of rules.

2.2   Quantitative Algorithm Ranking

The main objective of this study was to construct an algorithm that is applicable to binary classification tasks and evaluate it using datasets with various imbalance ratios. According to our review, researchers typically use the UCI database [13] to verify the accuracy of their algorithms. The problem is that in most of the corresponding datasets, only a small portion of the total data is provided; thus, all possible input instances are unavailable. Based on the incompleteness of these datasets, we may not be able to trust the results of the ranking of algorithms, even if we use different cross-validation methods, because we may inadvertently overfit or underfit the given data. To solve this problem, we aimed to generate datasets that can produce imbalance ratios similar to the UCI datasets while considering all possible data points, without noise (i.e., 100% of

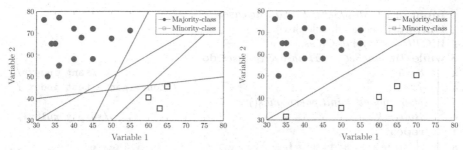

(a) Different solutions for defining a boundary

(b) Best boundary separator between two classes of data points

Fig. 1. Different data boundaries based on noise or new data entry.

possible instances for given inputs). We think that these artificial datasets, can give a good insight about real world because they cover different imbalance ratios via different linear and non-linear problems. Additionally, we intended to compare the ranking of algorithms using the generated data and UCI data.

Another important question is which metrics are the most suitable for ranking algorithms. Among the studies mentioned in the previous sections, most researchers used the accuracy metric to rank their algorithms. Unfortunately, this metric is not the most suitable for ranking algorithms when considering both balanced and imbalanced datasets. A few studies have used the area under the receiver operating characteristic curve (AUROC) as an evaluation metric. However, when faced with high imbalance ratios, the AUROC is not a suitable option. In such cases, one should consider using another metric such as the area under the precision-recall curve (AUPRC) [32]. In this study, we considered both the AUPRC and AUROC to rank different algorithms. By using relevant datasets and metrics, we identified the limitations of the AntMiner algorithms in terms of handling datasets with high imbalance ratios and attempted to overcome these limitations. To this end, we developed a novel algorithm called OCAntMiner, which is presented in Sect. 3.

3   OCAntMiner

This section presents our proposed OCAntMiner (One-Class AntMiner) algorithm. As shown in Fig. 1a, it is possible to define different boundaries for discriminating two classes of data, which becomes particularly important when dealing with imbalanced data.

The amount of data that can express a minority class is very small and noise in these data can completely distort the output of prediction. As shown in Fig. 1a, only three data points belong to square class (minority class) and all other data belong to class circle (majority class). Our goal in this study was to develop an algorithm that works on existing data without the need to add or remove data. As shown in Fig. 1b, if we add more data points to the existing data

Input   : *TrainingSet* : *all.training.cases*
Output: Discovered.Rules.list[]

1 $WCTP = MajorityClass$;
2 **while** *TrainingSet >Max uncovered cases* **do**
3 $\quad$ $t = 1$;   /*ant index*/
4 $\quad$ $j = 1$;   /*convergence test index*/
5 $\quad$ *pheromones = init.phermones*();
6 $\quad$ $Rule = []$;   /*empty rules*/
7 $\quad$ **repeat**
8 $\quad\quad$ Rule[t] = add terms based on heuristic function and pheromones;
9 $\quad\quad$ Prune Rule[t] ;   /*based on quality function*/
10 $\quad\quad$ **if** *Consequent of Rule[t] != WCTP* **then**
11 $\quad\quad$ $\quad$ Quality[Rule[t]]=0;
12 $\quad\quad$ **end**
13 $\quad\quad$ Update the pheromones;
14 $\quad\quad$ **if** *Rule[t] is equal to Rule[t-1]* **then**
15 $\quad\quad$ $\quad$ j = j + 1;
16 $\quad\quad$ **else**
17 $\quad\quad$ $\quad$ j = 1;
18 $\quad\quad$ **end**
19 $\quad\quad$ t = t + 1;
20 $\quad$ **until** *(t >= No.of.ants) OR (j >= No.rules.converged)*;
21 $\quad$ R.best =best(Rule) ;   /*Rule with highest quality among all Rules*/
22 $\quad$ **if** *Consequent of R.best == WCTP* **then**
23 $\quad\quad$ Add rule R.best to Discovered.Rules.List; ;
24 $\quad\quad$ TrainingSet=TrainingSet-(set of cases correctly covered by R.best);
25 $\quad$ **end**
26 **end**

Algorithm 1: High-level pseudocode for OCAntMiner

(either noise or true data) and if these data are added to the minority class, they can significantly change the boundaries of the minority class. However, for the majority class, the addition of new data does not significantly shift the boundary because the rest of the data can pinpoint the location of the boundary.

In previous versions of AntMiner, the majority class was considered as the default class and the algorithm searched for rules to explain the minority and majority classes with no restrictions on finding rules for each class. As a result, the algorithm could provide rules to describe the majority class and rules to describe the minority class, or only rules to describe the minority class. Our idea is to limit the algorithm to the majority class and extract rules only for that class. The reason for choosing the majority class to extract rules instead of the minority class is that there are more data for the majority classes in datasets, meaning more precise rules can be derived to express such classes. If we can describe one class very well, then another class will be easily discriminable. One goal of this study was to evaluate the impact of integrating this approach into AntMiner-based algorithms. The OCAntMiner pseudocode is presented in

Algorithm 1. OCAntMiner extends the original AntMiner by focusing on the extraction of the majority class, which is defined as the class with the highest frequency in the class distribution of training samples. The pheromone update, pheromone initialization, heuristic, and quality functions rely on the AntMiner model. To focus on the majority class, the first step is to detect which class to predict (WCTP), afterward, algorithm was modified to extract rules related to this class (the modifications are highlighted by red in Algorithm 1). Most other changes to the original version aim to prevent the algorithm from generating minority class rules. In line 10, the algorithm checks for the consequences of extracted rules. If the consequent does not match the value of the majority class, then the quality of the rule is equal to zero. Similarly, in line 22, when all ants provide their solutions (rules), the best solution (R.best) is selected based on the quality measure. If the consequent of R.best is equal to the value of the majority class, then R.best is added to the list of discovered rules and the cases correctly covered by R.best are removed from the training set.

4   Evaluation

In this section, we first provide a detailed overview of our datasets and then our methodology for evaluating algorithms. Subsequently, relevant evaluation metrics are identified and justified. Finally, the algorithms selected for the evaluation process are presented with their configurations.

4.1   Datasets

For evaluating and comparing their algorithms on classification tasks, most researchers use various UCI datasets [13] as a reference. However, as explained previously, these datasets do not provide all possible data instances, and consequently, we are unable to rely on ranking based on insufficient information. Additionally, we aimed to evaluate the algorithms under different conditions (i.e., different imbalance ratios), and it was unknown whether these datasets covered a large variety of possible scenarios.

To overcome these limitations and evaluate our algorithm, we used a dataset generator that allowed us to consider all of the complexities and relationships between input and output variables. Because we wished to generate datasets containing classes based on rules, we used binary logic to generate rules and considered inputs and outputs as binary values. The AND, OR, and NOT operators were used to generate all possible combinations and relationships between input variables. We assigned the desired number of input variables (four and eight in this study) to the generator for generating random rules. Then, based on the number of inputs, it generated all possible instances, and based on the generated rules, it evaluates the inputs and generated class outputs (e.g., for eight binary inputs, we generated $2^8 = 256$ different instances, that is, the number of possible instances without replacement for eight inputs).

Four and eight features as inputs may seem to be small numbers. However, as indicated in the results section, the complexity that these number of variables can generate is so high that there is no need to consider greater numbers of inputs at this time. The 24 generated datasets cover, imbalance ratios from 1 to 84. This method allowed us to generate all possible combinations of relationships between variables for each dataset and a very wide range of complexities. For assessment, we used five generated datasets with four inputs, 19 generated datasets with eight inputs, and five UCI binary datasets (Breast Cancer, Breast Cancer Wisconsin, Haberman, Hepatitis, and Tic Tac Toe).

4.2   Methodology

We first evaluated the algorithm performance using data that comes from data generator for the reasons mentioned in Sect. 4.1. The objective was to evaluate the performance of the algorithms in different scenarios. Therefore, we tested the algorithms with two different sampling percentages. In the first scenario, we fed all possible instances (i.e., 100% of the data) to the algorithms as training data, and use same data for testing, and then with metrics checked the behavior of the algorithms to determine which algorithms were capable of categorizing data in the best possible manner (i.e., fully predict the output and e.g. AUROC = 100%) considering the clarity and presence of answers in the data and the absence of noise. We know that this action may cause some models to overfited on the data. And we'll check that with the second scenario. Our objective in this scenario is just to verify the learning power of the algorithms. In the second scenario, we checked the performance of the algorithms using 50% of the all data points as training data and use 100% of data for testing. This scenario measured the robustness of the algorithms when only half of the samples were available for training. For this purpose, we randomly split the data into 10 folds (10-fold cross-validation) and, stored these folds. Then We run each algorithm 10 times and used 10 folds each time to make the results more reliable, and then use them to evaluate algorithms. Finally, we compare the ranking of the algorithms in two scenarios to see which algorithms were able to achieve the highest ranking in both scenarios.

To evaluate and understand the results of comparing the proposed algorithm to the other algorithms, we followed two steps. First, we described the results based on the mean and standard deviation of different runs. Second, statistical methods (i.e., AutoRank tools) were used to rank the algorithms based on their performance measures [16]. Specifically, we adopted the Demsar method [9]. The Demsar method was used to highlight both the ranking results of the benchmark and to show how much the performances of the different algorithms varied from each other. To compare the algorithms and rank them according to classification performance, because the datasets used for evaluation were highly imbalanced, we considered classification metrics that yield good results in this domain, including the AUPRC, AUROC [3], and Matthews correlation coefficient (MCC) [2].

4.3   Selected Algorithms

To demonstrate the performance of the proposed OCAntMiner algorithm compared to other algorithms, we selected algorithms from the category of evolutionary algorithms for the sake of fair comparison. However, we also went a step further and compared it to non-evolutionary algorithms to demonstrate the power of the proposed algorithm. We selected different versions of AntMiner, namely, the original version of AntMiner [18], cAntMinerPB [23], UCAntMinerPB [24], and cAntMiner [22]. For direct and non-evolutionary algorithms, we also selected some algorithms like RIPPER [7], PART [14], and RISE [11], and for indirect algorithms, we selected J48 [30,31], and REPTree [33].

4.4   Algorithm Parameters

We used the same input parameters for all algorithms, namely, "Size of Ant colony" = 60, "Maximum number of iterations" = 1500, "Minimum covered cases per rule" = 10, "Number of uncovered cases set" = 10, and "Rule quality function" = Sensitivity × Specificity. In the next section, we present the results of different algorithms based on the described metrics.

5   Results and Discussion

5.1   Evaluation of Algorithms from a Classification Perspective

In this section, we compare different algorithms from a classification perspective. For this purpose, we consider three different metrics (AUROC, AUPRC, and MCC) to evaluate the accuracy of dataset. To verify the algorithms in different scenarios, we applied two different sampling percentages (100% and 50%) to the input data. Our first scenario is to provide the algorithms 100% of the data samples (i.e., all possible instances without replication) for training and testing data to evaluate whether they can achieve the maximum value (e.g., AUROC = 100%) when all possible samples are used. Because all possible samples are provided to the algorithms, we expect the algorithms to adapt to the data and provide the maximum AUROC, AUPRC, and MCC values for the testing data. In Table 1, mean values of different runs were sorted by the AUPRC metric, and the highest values are bolded. In left side of this Table we consider a second scenario in order to see what happens when we give only half part of possible data samples, and which algorithm can predict the output better. As can be seen in the Table, the results decrease compared to using all possible data samples (right side of Table 1), but still, OCAntMiner gives the best result on AUROC, AUPRC, and MCC in comparison to other Antminer algorithms. However, when considering the non-evolutionary algorithms, RIPPER provides better performance in terms of the AUPRC metric.

Also, this Table shows results using 100% of all instances. Rise algorithm achieves the best values for all metrics but when we look at the result with 50%

Table 1. Mean [Std] using 24 synthetic datasets with different sampling.

	50% of whole possible instances			100% of whole possible instance		
	AUROC	AUPRC	MCC	AUROC	AUPRC	MCC
OCAntMiner	**90.64 [9.21]**	89.48 [16.66]	**80.82 [17.56]**	**95.10 [8.71]**	93.87 [11.37]	**92.72 [13.19]**
RIPPER	90.56 [15.59]	**92.43 [19.99]**	77.08 [32.84]	94.52 [14.13]	**95.34 [18.76]**	85.97 [30.68]
IREP	86.82 [17.28]	90.2 [20.77]	71.26 [35.33]	93.31 [14.58]	94.49 [19.01]	85.75 [29.09]
UCAntMinerPB	82.40 [18.35]	81.73 [24.73]	57.42 [35.23]	82.53 [21.81]	84.56 [24.66]	62.08 [42.44]
Jrip	81.58 [19.44]	87.42 [21.16]	63.40 [34.7]	81.57 [21.09]	88.90 [20.55]	66.03 [36.67]
cAntMinerPB	76.91 [19.1]	74.66 [29.49]	45.09 [33.78]	77.41 [19.25]	75.15 [28.83]	49.90 [35.52]
Rise	76.12 [5.12]	84.62 [12.99]	45.74 [20.58]	100.0 [0.0]	100.0 [0.0]	100.0 [0.0]
J48	72.42 [19.08]	81.39 [23.6]	43.39 [38.41]	78.71 [19.87]	87.13 [21.96]	53.67 [40.81]
AntMiner	72.32 [16.99]	73.69 [28.8]	41.70 [31.79]	75.02 [20.63]	74.84 [29.35]	48.12 [38.65]
REPTree	70.99 [19.33]	80.28 [23.35]	40.21 [38.54]	77.60 [19.67]	83.59 [23.12]	50.92 [39.27]
cAntMiner	69.45 [19.18]	73.83 [29.05]	36.11 [35.36]	73.54 [18.91]	74.78 [28.76]	45.82 [36.28]
PART	69.27 [18.59]	72.31 [30.19]	42.29 [33.42]	65.93 [21.22]	70.90 [30.91]	37.13 [37.4]

sampling, result dramatically decreases and it is sign of overfitting or underfitting. But the results of algorithms OCAntMiner and RIPPER have not differed much and have kept the same position as before. In Table 2, we compare the original AntMiner to the improved version (i.e., OCAntMiner) with more specific criteria using 50% sampling of all possible instances of the generated datasets. The results indicate that the number of rules is reduced by 43%, runtime is improved by 47%, and AUPRC is improved by 15.87% when using OCAntMiner. This demonstrates that the proposed algorithm provides fewer rules in less time and is more accurate for prediction. In this section, we presented the results in terms of the mean and standard deviation for different sampling percentages. In the next section, we examine the ranking of the algorithms in detail using statistical methods and the UCI datasets.

Table 2. Comparison using 50% of all possible instances of 24 synthetic datasets.

Algorithm	Number of rules	Time(S)	AUPRC
AntMiner	667	3400	78%
OCAntMiner	375 (↓ 43%)	1778 (↓ 47%)	93.87% (↑ 15.87%)

5.2 Ranking of Algorithms

In this section, we rank the algorithms using a statistical approach based on AutoRank tools using both UCI and synthetic datasets. Because the data under analysis do not follow a normal distribution, the Friedman test with the Nemenyi post-hoc test was applied to rank the algorithms and divide the algorithms into different groups based on the critical distance (CD) metric. As shown in Fig. 2, the algorithms were ranked based on the AUROC metric using UCI datasets. One can see that the most performant algorithm is the proposed OCAntMiner, followed by J48. This figure also shows that OCAntMiner, J48, RIPPER, and IREP are in the same group and there is no significant difference between them, but they are significantly different from the other algorithms. It also shows that the original version of AntMiner is in the group with the worst results along with different versions of AntMiner. With our modifications, it jumps to the first group and first rank. Additionally, we performed another test with 24 synthetic datasets and 50% random sampling. The results are presented in Fig. 3, and shown that, RIPPER has the highest ranking, followed by OCAntMiner. This figure also shows that OCAntMiner provides significantly better results than the other versions of AntMiner and is clearly in the top group. These results indicate that despite being in the same group as RIPPER, there is still room for improving OCAntMiner.

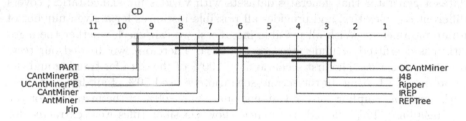

Fig. 2. AUPRC on UCI datasets with 10-fold cross-validation.

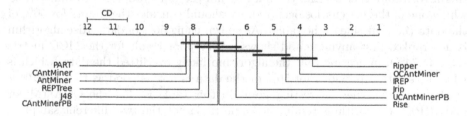

Fig. 3. AUPRC on synthetic datasets with 50% of data with 10-fold.

6   Discussion and Conclusions

In this study, we focused on rule-based classifiers, specifically the AntMiner algorithm family, for extracting classification rules from a given dataset. The most important part for the classification task are validation results, because if such results are not calculated properly, it can lead to incorrect directions for improving existing algorithms. For validation, there are two important features, the datasets, and metrics used for validation. Most studies have used the UCI database for evaluating algorithms in different areas with varying complexities. However, most UCI datasets cannot provide all possible instances for data inputs to facilitate the measurement of the intrinsic performance of algorithms when seeking a binary function. Interestingly, no AntMiner algorithm was able to consistently achieve values of 100% for the metrics used when all instances were provided. When evaluating and ranking different algorithms using synthetic datasets with 50% of all data instances, overfitting or underfitting to the given data may have occurred. The same phenomenon should occur with the UCI data, but because we did not have all data instances, we could not verify this phenomenon.

Another limitation of using UCI datasets is that we do not know the extent to which the data used for ranking cover different complexities for the target problem. Our contribution to solving these problems was the introduction of a dataset generator that generates datasets with various imbalance ratios, covers different complexities, and provides all possible instances for a given number of input parameters. With all possible instances, we can check whether the algorithm is overfitted or underfitted to the data. Therefore, we divided our tests into two scenarios. The first scenario used 100% of the data for training and the same data for testing. In the second scenario, we used 50% of all data instances with random sampling for the training phase and 100% of the instances for the testing phase. This allowed us to check how classifier rules were extracted for all data instances. The results of the different algorithms for 100% and 50% of all possible instances are presented in Table 1. As shown on the right side of this table (i.e., 100% sampling), the maximum values for AUROC, AUPRC, and MCC are 100% and the Rise algorithm can achieve these values, indicating that this algorithm either overfits the data or fits the data properly. To understand which one of these scenarios occurred, we should consider the results for 50% of the data (i.e., left side of the table). As shown using 50% of data, Rise algorithm is not ranked first anymore and shows a very poor result for the MCC metric (i.e., 45.74%), indicating that the algorithm likely overfitted the data, which is why the solution provided for 50% of the data is not as good as that for 100% of the data. Another interesting point in this table is that the OCAntMiner and RIPPER algorithms retain the same ranks for the two different sampling percentages based on different metrics, indicating that these two algorithms are robust to overfitting. This also indicates that they are likely not underfitted to the data because they achieve the highest values for different metrics.

Another issue that we observed in AntMiner was that this algorithm sets the majority class as the default class and then attempts to find rules to describe the majority and minority classes. As a result, three outputs may be generated at the end of executing the algorithm: 1) some rules describing the majority class and some rules describing the minority class, 2) all rules describing the minority class, or 3) no rules extracted from the data and only the default class is used for all data instances. As shown in Fig. 1, the distribution and boundary of majority data points are more reliable than those of minority data samples because there may be too few instances for describing the minority class. As a result, noise in the data or lack of instances may confuse the extraction of rules. This is why simply describing the majority class seems to be a powerful approach to solving this problem. Our main contribution in this study is highlighting the relevance of this approach, which, to the best of our knowledge, has not yet been applied to any AntMiner-based algorithm. We implemented this approach by modifying the first version of AntMiner and forced the algorithm to extract rules for the majority class alone. Therefore, the algorithm used the minority class as the default class and attempted to find the rules for majority data instances. As shown in Table 2, after adding this feature to the first version of AntMiner, the number of extracted rules decreased dramatically (43%) and the classifier with fewer rules could still detect the behavior of the data. This approach also reduced the runtime by 47%. Furthermore, the AUPRC was improved by 15.87%. These results demonstrate that with this added feature, we can reduce the runtime and number of rules while improving data classification performance. Additionally, as shown in Table 1, by adding this feature, we achieved a big jump in performance (from the second-worst AntMiner to the best). This jump demonstrates the strength of the components added to the algorithm to handle datasets with various imbalance ratios.

We performed a statistical test using the Demsar method to rank the algorithms considering the uncertainty in ranking caused by the number of datasets used in the experiments and variance of the results. As shown in Fig. 2, we first ranked the different algorithms based on the UCI datasets with 10-fold cross-validation to demonstrate how the proposed algorithm works on datasets used in previous studies. The results reveal that OCAntMiner ranks first among all evolutionary and non-evolutionary algorithms. Statistically, it is in the same group as the J48, RIPPER, and IREP algorithms. We also applied the Demsar method to the algorithms using 50% of all possible instances, and the results are presented in Fig. 3. The results show that OCAntMiner ranks second among all algorithms, but ranks first among the different versions of AntMiner. Additionally, this test indicated that OCAntMiner belongs to the same group as RIPPER and IREP. This figure also reveals that the ranking of the J48 algorithm is dramatically reduced compared to that in the previous figure using the UCI data. Several interpretations can explain this phenomenon. For example, synthetic datasets are more general than UCI datasets and cover more complex problems. Under these conditions, J48 may not classify some datasets properly or the five UCI datasets may not be sufficient to rank the algorithm (5% error in ranking). Finally, we demonstrated that modifying the AntMiner process can significantly improve its results without supplementing or modifying the heuristic or quality functions. In future work, we will consider adding this feature to other AntMiner models, and we also want to extend this method with multiple classes and analyzing the resulting algorithm behavior.

Acknowledgements. This research was conducted within the framework of the Offensive Sciences project number 13.11 "Virtual Innovative Real Time Factory"

(VIRTFac) which benefits from the financial support of the Offensive Sciences program of the Upper Rhine Trinational Metropolitan Region, the INTERREG V Upper Rhine program and the European Regional Development Fund (ERDF) of the European Union.

References

1. Agrawal, R., Srikant, R., others: Fast algorithms for mining association rules. In: Proceedings of the 20th International Conference Very Large Data Bases, VLDB, vol. 1215, pp. 487–499. Citeseer (1994)
2. Boughorbel, S., Jarray, F., El-Anbari, M.: Optimal classifier for imbalanced data using Matthews correlation coefficient metric. PLoS ONE **12**(6), e0177678 (2017)
3. Branco, P., Torgo, L., Ribeiro, R.: A survey of predictive modelling under imbalanced distributions. arXiv:1505.01658 [cs], http://arxiv.org/abs/1505.01658 (2015)
4. Bull, L., Kovacs, T.: Foundations of learning classifier systems: an introduction. In: Foundations of Learning Classifier Systems, pp. 1–17. Springer (2005). https://doi.org/10.1007/b100387
5. Chen, T., Guestrin, C.: XGBoost: A scalable tree boosting system. In: Proceedings of the 22nd ACM SIGKDD International Conference on Knowledge Discovery and Data Mining, pp. 785–794 (2016)
6. Clark, P., Niblett, T.: The CN2 induction algorithm. Mach. Learn. **3**(4), 261–283 (1989). DOI: https://doi.org/10.1007/BF00116835,http://link.springer.com/10.1007/BF00116835
7. Cohen, W.W.: Fast effective rule induction. In: Proceedings of the Twelfth International Conference on Machine Learning, Tahoe City, California, July 9–12, 1995, pp. 115–123 (1995)
8. Dargan, S., Kumar, M., Ayyagari, M.R., Kumar, G.: A survey of deep learning and its applications: a new paradigm to machine learning. Arch. Comput. Methods Eng. **27**(4), 1071–1092 (2019). https://doi.org/10.1007/s11831-019-09344-w
9. Demsar, J.: Statistical comparisons of classifiers over multiple data sets. J. Mach. Learn. Res. **7**, 30 (2006)
10. Dhaenens, C., Jourdan, L.: Metaheuristics for data mining: survey and opportunities for big data. Ann. Oper. Res. **314**(1), 117–140 (2022). https://doi.org/10.1007/s10479-021-04496-0
11. Domingos, P.: Unifying instance-based and rule-based induction. Mach. Learn. **24**(2), 141–168 (1996). https://doi.org/10.1007/BF00058656
12. Dorigo, M., Gambardella, L.: Ant colony system: a cooperative learning approach to the traveling salesman problem. IEEE Trans. Evol. Comput. **1**(1), 53–66 (1997). https://doi.org/10.1109/4235.585892
13. Dua, D., Graff, C.: UCI Machine Learning Repository (2017). http://archive.ics.uci.edu/ml
14. Frank, E., Witten, I.: Generating accurate rule sets without global optimization. In: Machine Learning: Proceedings of the Fifteenth International Conference (1998)
15. Ghannad, N., De Guio, R., Parrend, P.: Feature selection-based approach for generalized physical contradiction recognition. In: Cavallucci, D., Brad, S., Livotov, P. (eds.) TFC 2020. IAICT, vol. 597, pp. 321–339. Springer, Cham (2020). https://doi.org/10.1007/978-3-030-61295-5_26
16. Herbold, S.: Autorank: a Python package for automated ranking of classifiers. J. Open Source Softw. **5**(48), 2173 (2020). https://doi.org/10.21105/joss.02173

17. Holland, J.H.: Adaptation**research reported in this article was supported in part by the National Science Foundation under grant DCR 71–01997. In: ROSEN, R., SNELL, F.M. (eds.) Progress in Theoretical Biology, pp. 263–293. Academic Press (1976). https://doi.org/10.1016/B978-0-12-543104-0.50012-3, https://www.sciencedirect.com/science/article/pii/B9780125431040500123

18. Liu, B., Abbass, H., McKay, R.: Density-based heuristic for rule discovery with ant-miner. In: The 6th Australia-Japan Joint Workshop on Intelligent and Evolutionary System (2002)

19. Martens, D., De Backer, M., Haesen, R., Vanthienen, J., Snoeck, M., Baesens, B.: Classification with ant colony optimization. IEEE Trans. Evol. Comput. **11**(5), 651–665 (2007). https://doi.org/10.1109/TEVC.2006.890229, http://ieeexplore.ieee.org/document/4336122/

20. Medland, M., Otero, F.E.B., Freitas, A.A.: Improving the cAnt-MinerPB classification algorithm. In: Dorigo, M., et al. (eds.) ANTS 2012. LNCS, vol. 7461, pp. 73–84. Springer, Heidelberg (2012). https://doi.org/10.1007/978-3-642-32650-9_7

21. Otero, F.E.B., Freitas, A.A., Johnson, C.G.: cAnt-Miner: an ant colony classification algorithm to cope with continuous attributes. In: Dorigo, M., Birattari, M., Blum, C., Clerc, M., Stützle, T., Winfield, A.F.T. (eds.) ANTS 2008. LNCS, vol. 5217, pp. 48–59. Springer, Heidelberg (2008). https://doi.org/10.1007/978-3-540-87527-7_5

22. Otero, F.E.B., Freitas, A.A., Johnson, C.G.: Handling continuous attributes in ant colony classification algorithms. In: Proceedings of the 2009 IEEE Symposium on Computational Intelligence in Data Mining (CIDM 2009), pp. 225–231. IEEE (2009)

23. Otero, F.E.B., Freitas, A.A., Johnson, C.G.: A new sequential covering strategy for inducing classification rules with ant colony algorithms. IEEE Trans. Evol. Comput. **17**(1), 64–76 (2013). https://doi.org/10.1109/TEVC.2012.2185846

24. Otero, F.E.B., Freitas, A.A.: Improving the interpretability of classification rules discovered by an ant colony algorithm: extended results. Evol. Comput. **24**(3), 385–409 (Sep 2016). https://doi.org/10.1162/EVCO_a_00155, https://direct.mit.edu/evco/article/24/3/385-409/1025

25. Otero, F.E., Freitas, A.A.: Improving the interpretability of classification rules discovered by an ant colony algorithm. In: Proceeding of the Fifteenth Annual Conference on Genetic and Evolutionary Computation Conference - GECCO 2013, p. 73. ACM Press, Amsterdam (2013). https://doi.org/10.1145/2463372.2463382, http://dl.acm.org/citation.cfm?doid=2463372.2463382

26. Otero, F.E., Freitas, A.A., Johnson, C.G.: Inducing decision trees with an ant colony optimization algorithm. Appl. Soft Comput. **12**(11), 3615–3626 (2012). https://doi.org/10.1016/j.asoc.2012.05.028, https://linkinghub.elsevier.com/retrieve/pii/S1568494612002864

27. Parpinelli, R.S., Lopes, H.S., Freitas, A.: An ant colony based system for data mining: applications to medical data. In: Proceedings of the 3rd Annual Conference on Genetic and Evolutionary Computation, pp. 791–797. Citeseer (2001)

28. Parpinelli, R., Lopes, H., Freitas, A.: Data mining with an ant colony optimization algorithm. IEEE Trans. Evol. Comput. **6**(4), 321–332 (2002). https://doi.org/10.1109/TEVC.2002.802452, http://ieeexplore.ieee.org/document/1027744/

29. Pfahringer, B.: Random model trees: an effective and scalable regression method (2010)

30. Quinlan, J.R.: C4.5: Programs for Machine Learning. Morgan Kaufmann Publishers Inc., San Francisco (1993)

31. Quinlan, J.R.: C4.5: Programs for Machine Learning. Elsevier (2014)
32. Saito, T., Rehmsmeier, M.: The precision-recall plot is more informative than the ROC plot when evaluating binary classifiers on imbalanced datasets. PloS one **10**(3) (2015)
33. Srinivasan, D.B., Mekala, P.: Mining social networking data for classification using REPTree. Int. J. Adv. Res. Comput. Sci. Manage. Stud. **2**, 10 (2014)
34. Vikhar, P.A.: Evolutionary algorithms: a critical review and its future prospects. In: 2016 International Conference on Global Trends in Signal Processing, Information Computing and Communication (ICGTSPICC), pp. 261–265 (2016). https://doi.org/10.1109/ICGTSPICC.2016.7955308

Towards a Many-Objective Optimiser for University Course Timetabling

James Sakal(✉) [iD], Jonathan Fieldsend[iD], and Edward Keedwell[iD]

University of Exeter, Exeter, UK
{js1188,J.E.Fieldsend,E.C.Keedwell}@exeter.ac.uk

Abstract. The University Course Timetabling Problem is a combinatorial optimisation problem in which feasible assignments of lectures are sought. Weighted sums of violations of various constraints are used as a quality measure, with lower scores (costs) being more desirable. In this study, we develop a domain-specific many-objective optimiser, based on constructive heuristics and NSGA-III, in which the violations of different constraints are cast as separate objectives to be minimised concurrently. We show that feasible solutions can be attained consistently in a first phase and that a targeted objective can be fully optimised in a second phase. A set of non-dominated solutions is returned, representing a well-spread approximation to the Pareto front, from which a decision maker could ultimately choose according to *a posteriori* preferences.

Keywords: Many-objective · Optimisation · Timetabling

1 Introduction

The generalised University Course Timetabling Problem (UCTP) is the task of generating a workable university timetable by assigning lectures to discrete locations in time and space, subject to various constraints. It is a well studied problem in combinatorial optimisation and known to be computationally hard [19]. This study works with the standard curriculum-based formulation proposed by the International Timetabling Competition (ITC) 2007 Track 3 under the popular UD2 configuration [4,9]. While it is noted in [3] that all (unique) instances of this benchmark but 3 have been solved to optimality, this does not diminish its usefulness. The formulation remains challenging for optimisers running on short-to-medium timeouts, while prior knowledge of the optimal values helps to contextualise results. The reader is directed to the sources above for an in-depth description of the problem and constraints, which are modelled on the real world timetabling problem of the University of Udine. In brief, feasible timetable solutions cannot violate any of five given hard constraints **h1 ... h5**. These ensure that all lectures are assigned, pre-designated unavailable periods are avoided, as are clashes between lectures. The quality of a feasible solution is determined by violations of four soft constraints, **s1 ... s4**, which relate to room capacity, minimum working days, curriculum compactness and room consistency respectively.

P. Legrand et al. (Eds.): EA 2022, LNCS 14091, pp. 133–144, 2023.
https://doi.org/10.1007/978-3-031-42616-2_10

We use the following notation to refer to entities in the benchmark instances: $\mathcal{L}$ is the set of lectures $\{l_1 \ldots l_\gamma\}$, d_i a day of the week, t_i a timeslot within a day, $p_i = t \times d$ a period (or timeslot within a week), r_i a room. Adopting the terminology used in [16], a room/period pair is referred to as a *place*. This study proposes a parameterless many-objective optimiser based on the non-dominated sorting genetic algorithm III (NSGA-III) [6] and a constructive heuristic. The motivation is to evolve a set of solutions that approximate the Pareto front, thereby giving a decision maker a set of high quality timetables to select from. For efficiency, our approach incorporates δ-evaluators, as suggested by [12]. Phase 1 of the approach aims to find feasible starting solutions, which are then used to initialise the genetic algorithm in Phase 2. Here, the 4 soft constraint violation scores are cast as separate objectives to be minimised concurrently.

Section 2 provides some background work before Sect. 3 details the methodology and optimiser development. Section 4 describes the experiments and results. Sections 5 and 6 feature a discussion and conclusions respectively.

2   Related Work

While results have been published by many authors for the ITC2007 benchmark (see the Benchmark Analysis section of [13] for an incomplete list), the majority treat the problem as a single-objective minimisation, as prescribed by the original competition rules. The original Track 3 competition included five finalists [17], Z. Lu et al, [2,10] and [5], from which the multi-phase constraint-based solver of [17] was declared the winner. In the intervening years, the current best known single-objective results have been achieved by [1] and [15]. The former employed a hybrid genetic algorithm with Tabu Search, whose movement through the search space was determined by a sequence of large neighbourhood operators. The latter embedded an Adaptive Large Neighbourhood Search within a Simulated Annealing framework. The best known results are reproduced here for context.

It is noted in [14] that this single-objective approach predominates in educational scheduling generally, despite the existence of often numerous and conflicting objectives. The authors consider a 3–objective professional training scheduling problem with some similarities to the UCTP, comparing NSGA-II with NSGA-III. The former was found to be superior on all metrics except speed. However, the parameter values were tuned only for NSGA-II, and our problem has a higher-dimensional objective space which may be tackled better by NSGA-III. Other differences between the UCTP and the problem in [14] must be noted too, such as its timescale (repeating week-long blocks rather than months or years), requirement to assign all events, and lack of precedence constraints.

A more direct comparison may be made with [11], in which the many-objective nature of the UCTP and ITC2007 benchmark was considered. A trajectory search was carried out by selecting a small number of lectures and reassigning them. Various acceptance criteria were relied upon for the new evaluations. In both of the two approaches proposed, decision maker preferences were assumed *a priori* and implied by the cost function. This was defined as either the standard weighted sum of violations or the Chebyshev distance to a reference point (the origin). Using the latter resulted in a more even spread of scores across individual objectives.

To the best of our knowledge, there are as yet no published results for the benchmark that attempt to approximate the Pareto set in the absence of decision maker preferences. The following section outlines the development and reasoning behind the different components of our system.

3   Methodology

Encoding: Our system is built in MATLAB and incorporates modules from the platEMO optimisation suite [20]. Its first task is re-encoding the problem instances, by converting each problem from its original .ctt file format to a 2-D indexed cell array data structure.

Solutions to the problem — the timetables themselves — must also be encoded. This is a design choice with serious implications for the efficacy of any evolutionary algorithm used. The proposed solution encoding represents each assignment using the 3-tuple: $\langle d_i, t_i, r_i \rangle$, where d_i and t_i are the day and timeslot respectively and the element-wise length of a complete chromosome is $3 \times |\mathcal{L}|$. Disadvantages of using a 3-tuple include the larger data structure and higher time complexity involved, as well as the potential for epistatic effects caused by interactions between elements within tuples. More favourably, the induced search landscape grants connectivity between days, timeslots and rooms as individual entities, allowing for the design of more nuanced and effective genetic operators. Each element within a gene resonates with a particular soft constraint. For example, perturbing d_i affects the number of unique days that course lectures are held on, and therefore the violation score of **s2**. Compliance with **h1** (all lectures must be assigned) is also ensured by the 1:1 lecture:gene ratio.

Initialisation: The initialisation constitutes Phase 1 of a two-phase optimisation, with the aim being to produce a population of solutions that is as close to fully feasible as practicable. To this end, two broad categories of constructive heuristics have been proposed in the literature [18]. Static heuristics require lectures to be sorted by some metric, where this fixed ordering then determines the sequence of assignments. Dynamic heuristics involve recalculating the metric values after each assignment, thus providing greater adaptive potential. In both cases, the chosen metric is intended as a measure of 'difficulty to assign'.

The static heuristics Largest Enrolment (LE) and Largest Degree (LD) and the dynamic heuristic Saturation Degree (SD) were tested on the ITC2007 benchmark. LE relies on the number of enrolled students for its metric. Lectures with a larger number of students take priority. LD, as described for the generic case in [18], uses the number of potential clashes a lecture has with other lectures resulting from commonality of students. Since explicit student sectioning is not a feature of the ITC2007 benchmark, the metric is defined analogously as: The sum total of lectures that have either a curriculum or a teacher in common with

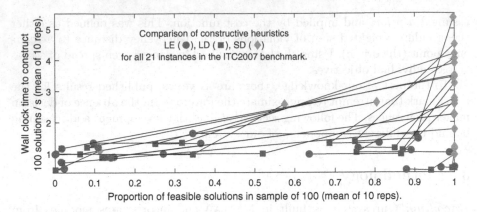

Fig. 1. Performance comparison of 3 constructive heuristics. Lines connect results for common instances.

the lecture being assessed. Priority is given to lectures with higher numbers of potential clashes in this respect. The metric for SD is the number of available feasible places, i.e. those that would not result in a hard constraint violation at the point of assignment. The lecture with the lowest value at each decision point is chosen for assignment. Across all heuristics, ties are broken at random.

Once a lecture has been chosen on the basis of its metric value, a place is randomly selected from the set of feasible places currently available to that lecture. If no feasible place exists, an infeasible place (excluding unavailable periods) is chosen at random instead. A secondary, period-based heuristic is suggested in [18] as an optional, more discriminatory, alternative to random sampling. Our system neglects to include this with the following justification: Any infeasible solutions that may have been constructed in Phase 1 are quickly bred out of the population by the inherent hard constraint handling mechanism. The extra expense of a period-based heuristic was therefore found to outweigh the marginal gains in feasibility rate.

In testing LE, LD and SD, 10 independent repetitions were carried out for each problem instance. In each repetition, 100 timetable solutions were constructed. This number was chosen to reflect the order of magnitude of a typical population. The primary quality measures to consider are the proportion of solutions that are feasible, and the relative speed of obtaining them. As with all experiments in this study, the computation was performed on a 12-core Ryzen9 with 32GB RAM, base clock speed 3.8GHz. The wall clock speed shown here resulted from using a single core and no parallelisation. Figure 1 shows the results for the three heuristics over the 21 instances.

SD achieves superior feasibility rates for every instance, while being computationally dearer. At the scale of a population size of 100, this additional time cost amounts to no more than a few seconds. More pertinently, all SD rates are 0.99 or higher, with the exception of the 3 instances comp02 (0.80), comp05 (0.11) and comp19 (0.58). Across the infeasible solutions constructed for these 3 problems, the mean distances to feasibility, given as a vector of the hard constraints (**h2 h3, h4, h5**), were (0.3, 3.1, 0.0, 0.2), (0.2, 18.6, 0.0, 0.2) and (0.5, 3.8, 0.0, 0.2) respectively. These show that in the minority of cases where SD fails to achieve a near-perfect feasibility rate, the expected violations of hard constraints in the infeasible subset are nonetheless low. In particular, **h4** is zero in all cases.

Besides feasibility and speed, there may be other factors to consider when assessing the quality of an initial population generated by a constructive heuristic. The percentage of unique individuals in the sample is one example. In the aforementioned tests, 100% was achieved across all instances and all heuristics on this measure. Additionally, it may be worth considering some measure of dispersion or dissimilarity between individuals. A suitably diverse starting population may be important in terms of the exploratory power of the optimiser.

Algorithm: NSGA-III is a successful evolutionary algorithm that supports many-objective optimisation with constraints [6]. It is an extension to the popular NSGA-II algorithm, which was originally conceived for lower-dimensional objective spaces [7]. As the ITC2007 problem has 4 objectives to optimise, NSGA-III serves as an appropriate base for Phase 2 of our system.

Selection and constraint handling: Alongside the initialised population, the SD heuristic implementation returns an array of feasibility flags, toggled during construction. The property `con` holds the flag associated with each solution, with a `true` value indicating at least one violation of a hard constraint. For the first generation only, scores for the four soft constraint objectives are then calculated in full. 2-way tournament selection is used to select a mating pool. Randomly paired candidate solutions are first compared on their `con` property, with the lower value indicating the winner. Feasible solutions are thereby given priority. Should the `con` values be equal, the sum of the objective scores is used as a tie-breaking fitness measure.

Genetic operators: For a real-valued encoding, NSGA-III traditionally uses simulated binary crossover (SBX) and polynomial mutation as its genetic operators. For this discrete problem, adaptations were first made to both genetic operators to ensure the preservation of integrality in decision space. Further investigation determined that, with no meaningful ordering apparent for entities such as days or periods, traditional polynomial mutation is not necessarily well suited for this problem domain. Similarly, standard SBX carries the risk of

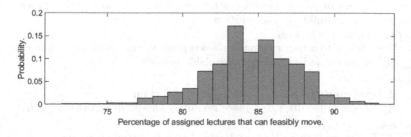

Fig. 2. A histogram of the percentage of assigned lectures with at least one feasible move available, for `comp12`. The sample set is 1000 feasible solutions constructed by SD.

degenerating timetables by recombining promising subsets in an injudicious way, thereby worsening the overall solution quality. In complex, combinatorial problems such as timetabling, a successful crossover operator requires domain-specific knowledge and can be computationally expensive. The proposed approach therefore dispenses with crossover entirely and is instead wholly reliant on a guided mutator. In developing this mutator, the following test was conducted:

1000 feasible solutions were constructed using SD. For each assigned lecture of each solution, a check was made on the number of places it could be re-assigned to without violating the overall solution feasibility. For some assignments, there were no feasibility-preserving moves available. The histogram in Fig. 2 shows an example (for comp12) of the distribution of percentages of assigned lectures, over the 1000 solution sample, with at least one such available move.

For all problems tested, the distributions demonstrate that the expected chance of an available feasible move is generally high. The optimiser can be guided, therefore, by imbuing the initial mutator, known as MuPF, with a preference for feasible moves where they exist. After randomly selecting one lecture, l_i, to be mutated, another random selection is made from the set of feasible moves available to that lecture. If this set is found to be empty, MuPF defaults the assignment to any random place.

Using this mutator, a test run was performed on comp01 with a population size of 364 over 550 generations. Over the course of this run, the minimum values of objectives (s1, s2, s3, s4) improved from (1599, 15, 88, 66) to (537, 0, 6, 28) respectively. Further tests emphasised the large relative contribution that s1 often makes to a scalarised objective score. An enhancement to the mutator, in which sufficient room capacity is considered, was proposed specifically to target this objective. Algorithm 1 outlines MuPFPR.

An initial indicative plot comparing MuPF and MuPFPR is given in Fig. 3. A run on comp01 was carried out with a function evaluation budget of 2 million. The

---

Algorithm 1: Preference for feasibility, preference for room (MuPFPR) mutation operator

---

Inputs: One starting solution
Output: One mutated solution
Randomly select a lecture, l_i, to mutate
Identify the set of places, feasMoves(l_i), to which l_i can be re-assigned without violating the feasibility of the solution
if $feasMoves(l_i) = \emptyset$ **then**
| Re-assign l_i to a new randomly chosen place in any room with sufficiently
|_ high capacity and excluding unavailable periods
else
| **if** $feasMoves(l_i) \cap sufficientRooms(l_i) = \emptyset$ **then**
| |_ Re-assign lecture i to a place randomly chosen from feasMoves(l_i)
| **else**
| | Re-assign l_i to a place randomly chosen from the given non-empty
|_ |_ intersection

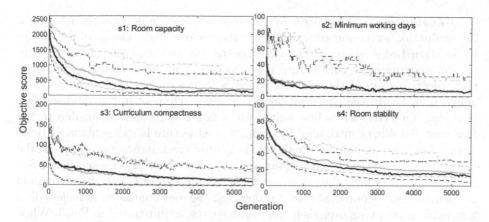

Fig. 3. A comparison of mutator MuPF (grey) and MuPFPR (black) for a single rep of comp01 with 2 million function evaluations. Traces shown are the min, mean and max objective scores over each generation.

extra room-related guidance provided by MuPFPR, shown as a black trace, helped drive the convergence rate for s1 objective in the top left tile, at no detriment to the remaining objectives.

Incorporated into the mutation process is an implicit feasibility checker. A violation flag, conMutation, is toggled if and only if feasMoves(i) = ∅. The returned con property for that child is generally given by (conParent ∨ conMutation) — except in the case when the parent solution is infeasible and the mutation is feasible. Here, the feasibility of the child is unknown and a full evaluation of the hard constraints must be called. The rarity of this outcome ensures that, in practise, the hard constraint evaluators seldom need to be executed at all — an example of a time-saving partial evaluation. The following section details how δ-evaluations are used to make similar savings when calculating the soft constraint objectives.

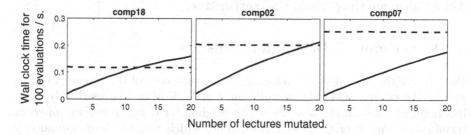

Fig. 4. A comparison of the time complexity (mean of 10 reps) for the combined δ-evaluators (solid line) vs. full (dashed line), for a small (comp01), medium (comp02) and large (comp07) sized problem and a variable number of mutations.

δ-evaluations: The process by which a δ-evaluation negates the need for a full evaluation on the soft constraint objectives is as follows: The ID of the lecture to be perturbed is recorded. The value contributed to the parent objective score by the assignment of this lecture is calculated. This value is subtracted from the objective score of that parent, which is known *a priori* from the previous generation. Lastly, the contribution of the new assignment in the child solution is added. Objective **s1** is best suited for a fast δ implementation, due to the fact that the value contributed by an individual lecture is independent of those from other lectures. For the remainder of the objectives, interactions between the lecture being perturbed and various other lectures must also be accounted for. Specifically, those from the same course (for **s2** and **s4**), or those with a common curriculum (**s3**). Combined over 4 objectives, the δ-evaluators nonetheless offer a sizeable time saving over their full counterparts, as illustrated in Fig. 4. While the run time of a full evaluator scales with the number of lectures, the δ run time scales with the number of mutations — due to the resulting combinatorial interactions. Under a single lecture mutation, the δ-evaluator gives the largest time savings, by multiples of 6.3, 10.7 and 13.2 for the respective problems shown.

Non-dominated sorting: NSGA-III relies initially on the dominance relation on objective scores to sort a concatenated parent/offspring population into non-dominated fronts. The efficient non-dominated sort with sequential search (ENS-SS) is used [21]. The hard constraint handling procedure mandates that any solution with a con flag value true is automatically dominated by all feasible solutions, regardless of the quality of its objective vector. The only way, therefore, in which such a solution can be admitted into the next generation is if the cardinality of the feasible solution set is less than the active population size. This in turn implies the following about Phase 2: If a given generation is fully feasible, all subsequent generations are also fully feasible. To promote diversity, NSGA-III also associates solutions with rays passing through a set of popSize uniformly distributed points on the 4-dimensional unit hyperplane. The normal-boundary intersection method with two layers is used to obtain these coordinates. popSize is a geometrically constrained approximation to the desired, user-input population size, setPopSize.

4   Experimental Design and Results

Each run of the optimiser was allocated to a single core of the Ryzen9 machine, as per the original ITC2007 stipulation. Parallelisation was used only across independent runs. In the absence of the original CPU benchmarking program, termination was after 600 s wall clock time, which was the limit intended by the competition, and setPopSize = 100. For each problem in a subset of 10 tested, 30 repetitions were carried out by varying the random seed. An external passive archive, implementing the ND-Tree structure [3,8], was constructed using the complete search history. The purpose was to update and store the set of non-dominated solutions found over the course of the search. The results are reported in terms of the following performance metrics: The best scalarised score found (using the original ITC2007 weighted penalty scheme). The size, at termination, of unique solutions in the non-dominated archive (both in decision and objective space, as the mapping is many-to-one). A Monte Carlo estimate of the hypervolume indicator, for which theoretical upper bounds on the maximum objective scores were used as the reference point coordinates.

Table 1 shows our results and statistics, alongside results from [1,11] and [15]. Figure 5 illustrates the spread of non-dominated solutions achieved by a single rep in 3-D objective space, for 3 problems in which the **s1** dimension has successfully been collapsed to zero.

Table 1. Results from 30 independent reps. b_s is the best scalarised solution score found over all reps, while $b_s($**s1, s2, s3, s4**$)$ gives the objective scores that make it up (averaged over the unique objective vectors whose sum is b_s). $\mathcal{A}$ is the final archive of non-dominated solutions, where sets of unique vectors in objective or decision space are distinguished by subscripts $_o$ and $_d$ respectively. Cardinalities for both are given as median values. $hv(\mathcal{A}_o)$ is the (mean) hypervolume of $\mathcal{A}_o$, while HV_{ref} is the reference point used. The best scalarised results from the two approaches in [11] are given as G1 (Threshold Accepting with 1% threshold) and G2 (reference point based). Finally, BK denotes the best known single-objective scores to date within the time limit, achieved by either [1]* or [15]† or both.

Instance	Proposed approach						Others		
	b_s	$b_s($**s1, s2, s3, s4**$)$	$\|\mathcal{A}_o\|$	$\|\mathcal{A}_d\|$	$hv(\mathcal{A})$	HV_{ref}	G1	G2	BK
comp01	11	(4, 0, 4, 3)	11	7492	0.959	(3606, 360, 294, 124)	5	10	5*†
comp03	162	(0, 52.5, 92, 17.5)	17	850	0.831	(11160, 720, 1536, 179)	115	154	68†
comp04	92	(0, 6.7, 65.3, 20)	17	482	0.853	(8151, 665, 1130, 207)	67	90	35*†
comp06	167	(0, 15, 104, 48)	16	233	0.777	(10632, 990, 1668, 253)	94	159	30*
comp08	108	(0, 0, 74, 34)	14	301	0.810	(7711, 700, 1166, 238)	75	120	37*
comp09	158	(0, 40, 94, 24)	24	623	0.821	(9269, 720, 1492, 203)	153	197	100†
comp11	0	(0, 0, 0, 0)	2	45453	0.981	(3196, 335, 500, 103)	0	0	0*†
comp13	131	(0, 30, 84, 17)	20	390	0.832	(10668, 670, 1292, 226)	101	133	59*†
comp14	125	(0, 20, 90, 15)	19	1289	0.866	(7138, 830, 1392, 190)	88	120	51†
comp18	116	(0, 30, 78, 8)	45	1373	0.884	(2638, 455, 954, 91)	n/a	n/a	64†

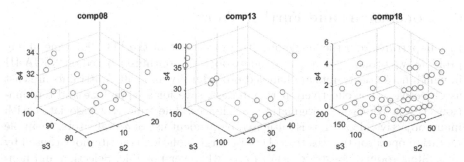

Fig. 5. Non-dominated solution sets in (s2,s3,s4)-space, found during single runs for 3 problems in which the fourth objective, s1, was optimised to zero.

5   Discussion

The strategy for speeding up (or by-passing) calculation of objective scores was successful in yielding inexpensive evaluations. However, this only partially mitigated against the cost of non-dominated sort. The algorithm unsurprisingly had a lower execution rate for function calls than many single-objective solvers. Comparing its performance on an equal function evaluation budget rather than a time budget would be enlightening, as the gradients in Fig. 3 suggest further gains are available. Despite this, scalarised results are seen to approach those of single-objective solvers on some problems which is encouraging — comp11 in particular was solved to optimality. With regard to the individual objective scores, the targeted operator MuPFPR was capable of rapidly optimising s1 to zero across the board (except for comp01 where the value of s1 in the optimal solution is known to be 4). These gains were not made at the expense of other objectives however, which showed improvement without exception during the runs. This suggests that additional bespoke operators, targeted at these objectives, may be a promising next step in striving to closer approximate the true Pareto front. A comparison with the reference point based approach of [11] (G2), shows competitive or improved scalarised scores, although this claim is weakened by the CPU benchmarking discrepancy. A major point of differentiation though is that our approach returns a population per run, rather than a single solution, in a comparable timescale. The approach appears relatively problem-agnostic, in contrast to [12] whose results show high variance across problems. Most importantly, it works on the assumption of *a posteriori* decision maker preferences. Different areas and extremes of the Pareto front are therefore explored simultaneously and a well-spread set of non-dominated solutions can be provided, as shown in Fig. 5. The hypervolume indicator values in Table 1 also evidence this, with all 10 problems, bar comp06, achieving a mean of 0.82 or higher. As lower absolute objective scores are achieved, the cardinality $|\mathcal{A}_o|$ naturally tends to decrease, as in comp01 (median 11) and comp11 (2). This can be explained by the proximity of the front to the origin and consequent sparsity of distinct points on the 4-D integer lattice. The observation $|\mathcal{A}_d| \gg |\mathcal{A}_o|$ also interestingly highlights the extent to which multiple designs map to a common objective point.

6   Conclusions and Further Work

In a departure from the single-objective treatment of the ITC2007 timetabling problem, we propose a two-phase, many-objective optimiser based on NSGA-III in which hard constraints are handled procedurally and soft constraints are cast as objectives. It is effectively paramaterless, save for setPopSize and termination criteria which are pragmatic user choices. The time cost associated with many-objective algorithms is mitigated by prudent use of δ-evaluators. A simple mutation operator reduces the otherwise large violation contributions caused by over-filling rooms (constraint s1) to zero wherever possible. Selection and non-dominated sorting ensure convergence of the other objectives as well as feasibility of solutions, while a quick start is guaranteed by the SD constructive heuristic.

Further work will focus on increasing the convergence speed of the remaining 3 objectives by widening the pool of targeted operators. If the mutator is considered as a neighbourhood, a more systematic exploration may be possible. Figure 2 gives an intuition about the size of such a neighbourhood. An adaptive element may be added to Phase 2 to select from such a pool based on the state of the current population or trajectory of the evolution. Alternatively, objectives that reach optimality may be aggregated with con so that any solutions sub-optimal in this objective will thereafter be automatically dominated. Further analysis will also help characterise the trade-offs between the objectives. By their definitions, $s1/s4$ and $s2/s3$ represent the two pairs with the greatest potential to conflict. The large cardinalities of the decision space solution sets suggests that genotype diversity could also play a useful role in the selection process.

References

1. Abdullah, S., Turabieh, H.: On the use of multi neighbourhood structures within a tabu-based memetic approach to university timetabling problems. Inf. Sci. **191**, 146–168 (2012)
2. Atsuta, M., Nonobe, K., Ibaraki, T.: ITC-2007 Track 2: An approach using general CSP solver. In: Proceedings of the Practice and Theory of Automated Timetabling (2007)
3. Bagger, N.C.F., Sørensen, M., Stidsen, T.R.: Dantzig-Wolfe decomposition of the daily course pattern formulation for curriculum-based course timetabling. Eur. J. Oper. Res. **272**(2), 430–446 (2019)
4. Bonutti, A., De Cesco, F., Di Gaspero, L., Schaerf, A.: Benchmarking curriculum-based course timetabling: formulations, data formats, instances, validation, visualization, and results. Ann. Oper. Res. **194**(1), 59–70 (2012)
5. Clark, M., Henz, M., Love, B.: QuikFix a repair-based timetable solver. In: 7th International Conference on the Practice and Theory of Automated Timetabling, PATAT 2008 (2008)
6. Deb, K., Jain, H.: An evolutionary many-objective optimization algorithm using reference-point based non-dominated sorting approach, part i: solving problems with box constraints. IEEE Trans. Evol. Comput. **18**(4), 577–601 (2013)
7. Deb, K., Pratap, A., Agarwal, S., Meyarivan, T.: A fast and elitist multiobjective genetic algorithm: NSGA-II. IEEE Trans. Evol. Comput. **6**(2), 182–197 (2002)
8. Fieldsend, J.E.: Data structures for non-dominated sets: implementations and empirical assessment of two decades of advances. In: GECCO 2020 - Proceedings of the 2020 Genetic and Evolutionary Computation Conference, pp. 489–497 (2020)
9. di Gaspero, L., Schaerf, A., McCollum, B.: The second international timetabling competition: curriculum-based course timetabling (Track 3). In: Proceedings of the 1st International Workshop on Scheduling a Scheduling Competition (2007)
10. Geiger, M.J.: An application of the threshold accepting metaheuristic for curriculum based course timetabling. In: Proceedings of the 7th International Conference on the Practice and Theory of Automated Timetabling (PATAT) (2008)

11. Geiger, M.J.: Multi-criteria curriculum-based course timetabling—a comparison of a weighted sum and a reference point based approach. In: Ehrgott, M., Fonseca, C.M., Gandibleux, X., Hao, J.-K., Sevaux, M. (eds.) EMO 2009. LNCS, vol. 5467, pp. 290–304. Springer, Heidelberg (2009). https://doi.org/10.1007/978-3-642-01020-0_25

12. Geiger, M.J.: Applying the threshold accepting metaheuristic to curriculum based course timetabling. Ann. Oper. Res. **194**(1), 189–202 (2012)

13. Gozali, A.A., Fujimura, S.: Solving university course timetabling problem using multi-depth genetic algorithm. SHS Web Conf. **77**, 01001 (2020)

14. Hafsa, M., Wattebled, P., Jacques, J., Jourdan, L.: A Multi-objective evolutionary approach to professional course timetabling: a real-world case study. In: 2021 IEEE Congress on Evolutionary Computation, pp. 997–1004 (2021)

15. Kiefer, A., Hartl, R.F., Schnell, A.: Adaptive large neighborhood search for the curriculum-based course timetabling problem. Ann. Oper. Res. **252**(2), 255–282 (2017)

16. Lewis, R., Paechter, B., Rossi-Doria, O.: Metaheuristics for university course timetabling. Stud. Comput. Intell. **49**, 237–272 (2007)

17. Müller, T.: ITC2007 solver description: a hybrid approach. Ann. Oper. Res. **172**(1), 429–446 (2009)

18. Pillay, N., Özcan, E.: Automated generation of constructive ordering heuristics for educational timetabling. Ann. Oper. Res. **275**(1), 181–208 (2019)

19. Rossi-Doria, O., et al.: A comparison of the performance of different metaheuristics on the timetabling problem. Pract. Theory Autom. Timetabling IV **2740**, 329–351 (2003)

20. Tian, Y., Cheng, R., Zhang, X., Jin, Y.: PlatEMO: A MATLAB platform for evolutionary multi-objective optimization. IEEE Comput. Intell. Mag. **12**(4), 73–87 (2017)

21. Zhang, X., Tian, Y., Cheng, R., Jin, Y.: An efficient approach to nondominated sorting for evolutionary multiobjective optimization. IEEE Trans. Evol. Comput. **19**(2), 201–213 (2015)

Empirical Investigation of MOEAs for Multi-objective Design of Experiments

Alexander Evans and Tinkle Chugh[✉][iD]

Department of Computer Science, University of Exeter, Exeter, UK
{aje220,t.chugh}@exeter.ac.uk

Abstract. Many machine learning algorithms require the use of good quality experimental designs to maximise the information available to the model. Various methods to create experimental designs exist, but the solutions can be sub-optimal or computationally inefficient. Multi-objective evolutionary algorithms (MOEAs), with their advantages of being able to solve a variety of problems, are a good method of creating designs. However, with such a variety of MOEAs available, it is important to know which MOEA performs best at optimising experimental designs. In this paper, we formulate experimental design creation as a multi-objective optimisation problem. We compare the performance of different MOEAs on a variety of experimental design optimisation problems, including a real-world case study. Our results show that NSGA-II can often perform better than NSGA-III in many-objective optimisation problems; RVEA performs very well; results suggest that using more objectives can create better quality designs. This knowledge allows us to make more informed decisions about how to use MOEAs when creating metamodels.

Keywords: Pareto optimality · Metamodelling · Evolutionary Computation

1 Introduction

Computer simulations are widely used in many scientific fields to understand systems that are complex or difficult to measure in the real world. Problems arise when simulations become computationally expensive. If one wants to understand the landscape, a small set of samples can be used to construct a metamodel. A metamodel is a regression model representative of a simulator. This allows the prediction of unsimulated areas of the landscape without expensive simulator runs. The problem of metamodeling and experimental designs is to determine what values to run the true simulator so that the metamodel regression is as accurate as possible [8]. Intuitively, it is best to uniformly spread the sample points across the domain, to maximise the information available for the regression metamodel. Uniform spread, or space filling, is the main concern of creating experimental designs; how do we position the sample points used for the metamodel across the domain space?

P. Legrand et al. (Eds.): EA 2022, LNCS 14091, pp. 145–158, 2023.
https://doi.org/10.1007/978-3-031-42616-2_11

There are various ways to create experimental designs, the most simple method is random sampling/Monte Carlo sampling. This is very limited in its use for metamodeling, as many samples are required to fill space effectively [8]. Latin hypercube sampling (LHS) improves on random sampling by considering a one-dimensional projection property for all sample points. LHS, when combined with space filling criteria, can create effective space filling designs; however, maintaining the one-dimensional projection property is difficult, as it is a strict constraint. Methods to obtain optimised LHS are computationally expensive and for some design parameters become infeasible [15]. Single objective methods return single solutions; no alternatives are given.

By employing multi-objective optimisation (MOO) in the creation of design of experiments (DOE), we can overcome these issues and give the decision maker (DM) greater control over the optimisation process. Multiple desirable properties of experimental designs can be chosen by the DM and constraints upon solutions can be applied [6]. For example, we could set a constraint that requires solutions to be Latin hypercubes/maintain single-dimensional projection. Alternatively, the single dimension projection ability of a design can be measured as an objective that is optimised in conjunction with other objectives; this may not give exact LHSs, but it can produce families of designs close to pure LHSs in a fraction of the time.

In addition to speed and customisability, MOO facilitates the creation of a set of optimal solutions, which provides many alternatives with different evaluation values [6]; the DM can select a design that fits his/her requirements. For experimental designs specifically, the presence of alternatives is especially powerful due to the multi-modality of the problem. MOO of DoE is a multi-modal multi-objective optimisation problem (MMOP). As such, experimental designs with similar evaluation values can have vastly different sample point locations. This gives the DM an even greater choice [18]. If, for example, a chosen design produces a substandard metamodel, the decision maker has not to change his requirements; s(he) can select another experimental design that is similar within the objective space and distant in the solution space [18]. This new design still meets the decision makers requirements however may produce a far better metamodel.

With the advantages of customisability and easy access to alternatives, the use of MOO for the creation of experimental designs is considered appropriate and should be explored. MOO is frequently done with the use of multi-objective evolutionary algorithms (MOEAs); these algorithms have various strengths and weaknesses. MOEAs can solve many types of problems; they can solve non-convex problems and without derivatives [6]. They are a good choice for solving the problem of design of experiments; however, they must be prepared to overcome the unique problems presented by multi-objective design of experiments. These problems include:

1. Large Gene Count: Due to the encoding methods, each potential solution in a modest DOE optimisation problem can contain hundreds of genes. As the

number of genes becomes very large the search space increases and algorithm performance deteriorates [19].

2. Multi-Modality: Although multi-modality can be advantageous, it comes with some drawbacks. For multi-modal problems, diversity management subroutines in MOEAs can inadvertently reduce diversity in the population and therefore the solution set [18].

3. Many-Objectives: A multi-objective problem with more than three objectives is called a many-objective problem. When the number of objectives increases, the effect of evolutionary operators on the population deteriorates and algorithms can struggle to converge on the optimum [12]. In our experiments, we are executing multi and many objective problems; the algorithms must be equipped to handle both.

To understand how to best use MOO for the creation of experimental designs, we will evaluate the performance of different MOEAs in their creation. By comparing performance, we can in the future select the correct algorithms to overcome the challenges of MOO of DOE, and fully reap its benefits. Furthermore, research into how the number of objectives affects design quality has not been explored. By performing experiments on different numbers of objectives, we can understand how adding more objectives affects the quality of the designs.

The rest of the article is structured as follows. In Sect. 2, we provide a background of MOO and DOE. In Sect. 3, we formulate the DOE as a multi-objective optimisation problem. In Sects. 4 and 5, we provide results for several benchmark and real-world problems by using different MOEAs. Finally, we conclude and mention the future research directions in Sect. 6.

2 Background

Criteria for space filling are widely researched in the experimental design field. They can be defined via distance based criteria, for example, minimax, maximin [13], potential energy [2]; or uniformity based criteria, where deviation from a uniform distribution is measured. More obscure criteria include correlation based and collapsibility criteria. Often, a single criterion is selected to optimise the sample points in an experimental design. We can remove this consideration and consider multiple objectives to create designs via multi-objective optimisation. We consider MOPs of the following form:

$$\text{minimize } \mathbf{f} = \{f_1(\mathbf{x}), \dots, f_k(\mathbf{x})\} \text{ subject to } \mathbf{x} \in S, \tag{1}$$

with k (≥ 2) objective functions and the feasible set S is a subset of the decision space $\Re^D$. A solution $\mathbf{x}^1$ dominates another solution $\mathbf{x}^2$ if $f_i(\mathbf{x}^1) \leq f_i(\mathbf{x}^2)$ for all $i = 1, \dots, k$ and $f_i(\mathbf{x}^1) < f_i(\mathbf{x}^2)$ for at least one $i = 1, \dots, k$. If a solution is not dominated by any of the possible solutions, it is called non-dominated. The set of such solutions is called the Pareto set. The aim of solving MOP is to find an approximated set of Pareto optimal solutions.

There are various methods of multi-objective optimisation: weighted sum, lexicographic ordering, and multi-objective evolutionary algorithms. All have been used to optimise experimental designs with promising results. In [14], multi-objective designs were created by combining the maximin and linear correlation criteria. Their designs were good, however, their use of weighted sums makes their results weaker, as weighted sum requires strong consideration of user preference and leaves results open to human error. Moreover, the weighted sum approach is not suitable for non-convex problems [16]. Abdellatif et al. [1] use lexicographical ordering to create hybrid Latin hypercube designs that optimise both the maximin criterion and the orthogonality criterion. Although they considered the proper order of optimisation, lexicographical ordering has weaknesses concerning the limitation of the search space. Gunpinar [9] used a multi-objective approach to create a genetic algorithm selection technique for computer-assisted design. Li et al. [15] created designs using the potential energy and maximin criteria to optimise designs via a modified NSGA-III. They did not consider the use of other algorithms. We will build upon their work by investigating which MOEAs are best for optimising experimental designs.

MOEAs attempt to find a evenly distributed approximation of the Pareto-optimal set of solutions. They use evolutionary operators like crossover, mutation, and selection to converge on a global optimum. In lower dimensional spaces where the Pareto set is one or two dimensions finding the optimal set is simple. Algorithms like NSGA-II [6] can perform very well at these tasks; however, as the number of objectives increases, selection pressure falls and convergence upon the optimum is weakened [12].

Work has been done to combine decomposition with Pareto-based approaches. NSGA-III uses predefined reference points. Reference points help select solutions from the non-dominated set, maintain diversity, and enhance convergence. These reference points must be chosen by the user although typically are uniformly distributed. NSGA-III selects members that are non-dominated and close to the given reference points. Proposed by Deb and Jain [7], they showed that NSGA-III produces good results for problems of up to fifteen objectives.

RVEA [4] also uses reference points to guide selection. Like NSGA-III, RVEA partitions the objective space, and selection is performed individually inside each partition. This helps balance diversity and convergence. The authors of RVEA showed that RVEA is a competitive algorithm when compared to NSGA-III; in some test problems it outperformed.

Indicator-based approaches, like Indicator Based EA (IBEA) [20], don't use dominance as selection measure but a user specified indicator. Indicators include hypervolume or eta indicators. Therefore, indicator-based approaches do not suffer the issues of dominance-based evolutionary algorithms. They can be prohibitively expensive when the number of objectives is too large [4]. NSGA-II, NSGA-III, RVEA, and IBEA are the algorithms that we shall use for the construction of designs. These algorithms have been chosen because they are commonly used and cover various paradigms of algorithm design.

3   Multi-objective Design of Experiments

In this section, we define the objective functions and formulate the design of experiment as a multi-objective optimisation problem.

3.1   Objective Functions

We have chosen four criteria that are appropriate for a design. All four are to be minimised. Having a selection of four different criteria allows evaluation of performance for different numbers of objectives. We can test the performance of each algorithm by constructing designs via two, three, and four objectives.

Potential Energy (AE). A popular space filling criterion, the Audze-Eiglais criterion [2] (also known as the potential energy criterion) fills space by treating each design point as a charged particle that repels all other particles. The total potential energy between the particles is used to evaluate their space filling. A design with low potential energy suggests the particles are spread uniformly across the domain. We chose this criterion for its excellent space filling properties. The potential energy criterion, for a design X_N, where N is the total number of samples, is denoted as:

$$PE(X_N) = \sum_{n=1}^{N-1} \sum_{j=n+1}^{N} \frac{1}{dis(\mathbf{x}^n, \mathbf{x}^j)},$$

where $dis(\mathbf{x}^n, \mathbf{x}^j)$ is the Euclidean distance between $\mathbf{x}^n$ and $\mathbf{x}^j$.

L2. Derived by Hickernell [10], the centred L2 discrepancy criterion assesses space filling by quantifying the distance between the continuous distribution of the design points and a discrete uniform distribution. We chose this criterion because it is also an effective space filling criterion that optimises from a different perspective to potential energy. For design X_N^D; where N is the number of sample points, D is the number of dimensions, and x_d^n is the nth sample in dimension d, the metric can be denoted as:

$$L_2(X_N^D) = \left(\frac{13}{12}\right)^D - \frac{2}{N}\sum_{n=1}^{N}\prod_{d=1}^{D}\left(1 + \frac{1}{2}|x_d^n - 0.5| - \frac{1}{2}|x_d^n - 0.5|^2\right)$$

$$+\frac{1}{N^2}\sum_{j,n=1,j\neq n}^{N}\prod_{d=1}^{D}\left(1 + \frac{1}{2}|x_d^n - 0.5| + \frac{1}{2}\left|x_d^j - 0.5\right| - \frac{1}{2}\left|x_d^n - x_d^j\right|\right)$$

Collapsibility (Coll). The non-collapsibility of a Latin hypercube is advantageous for an experimental design. When two points do not have a mutual coordinate they are said to be non-collapsible. A design is non-collapsible when no two points lie along the same one-dimensional slice; no two points share the same coordinate. Having non-collapsible points can save resources and provide more information per simulation run. Suppose that two points are collapsible along a single coordinate/variable; that is, they have the same or very similar value. If another variable/coordinate value has very little impact on the output of the simulator, those two design points will give similar outputs with no further information gained. Therefore, minimising the collapsibility of a design is important for its effectiveness; we have chosen to use a collapsibility criterion for the optimisation.

Collapsibility does not guarantee an effective space filling design; using this criterion in conjunction with other space filling criteria will allow its advantages to be fully utilised. Using the formula below can only be done using a multi-objective technique; by itself it is useless for space filling. Bates et al. [3] discussed this penalisation method that allows me to assess collapsibility.

We can assess collapsibility by evaluating each one dimensional projection of the sample points. If we take the dth-coordinate of all sample points in a design and sort them from smallest to largest we get the set $M_d = \{m_{d1}, m_{d2}...m_{dn}\}$. We can then create a set of equally spaced intervals that each point in M_d should lie appropriately within, $L = \{l_1...l_x\}$; where $x = N + 1$, l_1 is minimum of the sample space, and l_x is the maximum of the sample space. For a design to be a true Latin hypercube, each m_{dn} should lie within the interval $l_n \leq m_{dn} \leq l_{n+1}$. We check this equality across every m_{dn}, if any conflicts occur, we penalise the design. For a design, we sum the number of conflicts across all dimensions. A design with no conflicts is a Latin hypercube and the function would return zero. The function treats collapsibility as a minimisation problem. For a design X_N^D we can write the function as:

$$Coll(X_N^D) = \sum_{d=1}^{D} \sum_{n=1}^{N} A(M_{dn}), \quad A(M) = \begin{cases} 0, & \text{if } l_n \leq M \leq l_{n+1} \\ 1, & \text{otherwise} \end{cases}$$

Correlation (Corr). A design that has a strong correlation between its points will have areas of the domain space unexplored, which is undesirable. However, a design that has a low correlation is not guaranteed to be space filling. Using the correlation criterion in conjunction with space filling criteria ensures that the design is non-correlated and also space filling. By including this criterion the quality of the designs should increase. In our work we shall be using the Pearson coefficient; we try to minimise the largest pairwise correlation found across the design points. If R_X is the Pearson correlation matrix of each point in design X and I is an identity matrix of the same size, we can evaluate correlation in a single value denoted as:

$$Corr(X) = \max |R_X - I|$$

3.2   Encoding

For evolutionary algorithms encoding must be considered. If we consider an experimental design to be a system of N coordinates in an D dimensional hypercube we can represent a design as a N by D array. Most MOEAs do not support manipulating multi-dimensional arrays within their evolutionary operators; therefore, conversion is required. When we perform evolutionary operations upon each individual we flatten the multidimensional array into a single one dimensional array. When we evaluate the performance of each solution/design we reshape the one dimensional array into its true N by D array.

Each solution is represented as an array of length ND, where every component of each coordinate is a gene that can be operated against. Each gene is a real number between 0 and 1; this is done for ease of optimisation. For example, selecting 10 samples for a 5 dimensional simulator will grant me 50 genes per potential solution. The magnitude of samples can increase quite dramatically, for 200 samples in 5 dimensions the number of genes is 1000 per solution.

Bates et al. [3] compared our encoding solution to an alternative, where each sample point is represented as a single node number in the design space. The design contains a finite number of nodes each represented by an integer. A design can be represented by a sequence of integers each representing the nodes at which each sample is placed. We will not be using this encoding system as Bates et al. explains; the coordinates based encoding system requires less bits and therefore has a lower risk of encountering numerical errors.

4   Numerical Experiments

In this section, we compare different MOEAs with different combinations of objectives defined in the previous section.

4.1   Problem Specifications and Mumerical Settings

To test the limits of the MOEAs, several experiments with different parameters shall be executed - each building on the previous. The table below describes the specifications of each problem.

Experiment	Samples	Dimensions	Genes	Objectives
DOE 5.2	25	5	125	AE, Coll
DOE 5.3	25	5	125	AE, Coll, L2
DOE 5.4	25	5	125	AE, Coll, L2, Corr
DOE 10.2	50	10	500	AE, Coll
DOE 10.3	50	10	500	AE, Coll, L2
DOE 10.4	50	10	500	AE, Coll, L2, Corr
DOE 25.2	40	25	1000	AE, Coll
DOE 25.3	40	25	1000	AE, Coll, L2
DOE 25.4	40	25	1000	AE, Coll, L2, Corr

The experiment names are based on the parameters; a suffix of "5.2" refers to a 5 dimensional design optimised by 2 objectives.

Hypervolume shall be used as a performance measure upon the algorithms NSGA-II, NSGA-III, IBEA, RVEA. The reference point is constant across problems with a mutual number of objectives; for two objectives it is 1500, 1000; for three objectives it is 1500, 1000, 100; for four objectives it is 1500, 1000, 100, 2. RVEA parameters include an adaptation frequency of 0.2 and a rate of change of penalty of 2. IBEA used a kappa value of 0.05. Simulated binary crossover and polynomial mutation were used, both with a distribution index of 20 and a probability of 1. Initial population size of 200; the initial population is identical across problems with mutual levels of dimensionality. Termination occurs after 100,000 function evaluations.

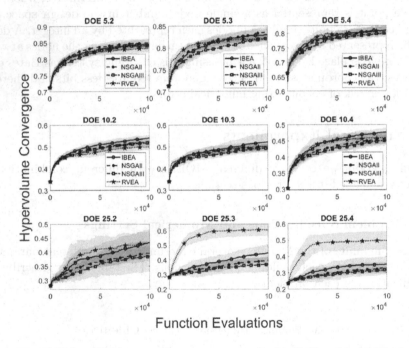

Fig. 1. PlatEMO hyper-volume performance across all nine DoE problems. The legend is the same for all subplots.

4.2   Results and Discussion

The results for hypervolume convergence can be seen in Fig. 1. NSGA-III is designed for many objective problems therefore it is expected to perform better than NSGA-II in 4 objectives [7], however the results suggest otherwise. Ishibuchi et al. [11] showed research that suggests that the choice of problem has a larger effect on performance comparisons than the number of objectives.

In their algorithm evaluation of a 500 item knapsack problem they showed that NSGA-II performs consistently better up to 10 objectives. DOE problems and knapsack problems are similar in that each individual is represented by a large number of genes. The performance of NSGA-II over NSGA-III remains constant across all hypervolume convergence graphs in Fig. 1; the large number of genes in DoE problems could be a factor in explaining the results. Ishibushi et al. also showed that NSGA-II performs better than NSGA-III when the Pareto front is very large compared to the spread of the initial solutions. For these problems, strong diversification is needed [11]. Figure 2 shows the initial population and the final population for NSGA-II and NSGA-III, we can see that the difference in spread between the final and initial populations is large and that NSGA-II produces a more diverse final population. NSGA-II's crowding distance diversity measure seems to perform better on this class of problem, as it does with knapsack problems.

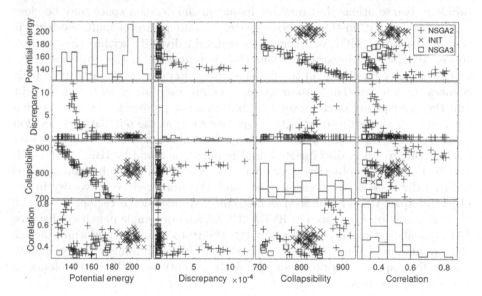

Fig. 2. DOE 25.4: NSGA-II and, NSGA-III final and initial population.

As both algorithms had the same initial population, it would be worth confirming performance comparisons by re-running the experiment with a different initial population. Different methods of initial population generation should be considered also; in these experiments initial populations were random. Perhaps an initial population of Latin hypercubes would produce better results as the collapsibility criterion is attempting to achieve Latin hypercube qualities. A non-optimal Latin hypercube initial population would help the MOEAs produce good designs with less work.

RVEA performed better than other algorithms for problems with more than two objectives. Cheng et al. compared RVEA with other popular MOEAs and showed their performance was better than other many-objective evolutionary algorithms. RVEA's strengths in benchmark problems have been replicated in MOO of DoE. This high performance is likely due to the unique scalarisation approach employed by RVEA. IBEA has consistently good performance across all test problems, this suggests it is good as a general use algorithm for MOO of DoE.

Cheng et al. [5] showed that in many objective problems with high gene count, IBEA and RVEA performed better in approximating the Pareto front than NSGA-III. DoE MOO's high gene count has replicated these results as both IBEA and RVEA perform better than NSGA-III across all problems. However, in Cheng et al.'s work neither IBEA nor RVEA perform best overall, which is also confirmed by our hypervolume results.

Design of experiments MOO is a multi-modal multi-objective optimisation problem, two solutions that may be distant in the decision space may be close or overlapping within the objective space. A consequence of multi-modality is that conventional MOEAs struggle to maintain diversity within the decision space. MOEAs will remove solutions that are crowded in the objective space when they may be distant in the decision space. Removal of distant individuals reduces diversity in the decision space. This process - along with genetic drift and the consequence of crossover and mutation not producing diverse offspring effectively - reduces diversity in the objective space as the population's decision variables are somewhat homogeneous [18]. Consequences of multi-modality may explain the irregular, disconnected final populations found by IBEA, NSGA-II, and NSGA-III; the objective space can be seen in Fig. 3. Multi-modality has reduced diversity in the decision space and, therefore, reduced diversity in the objective space that can be seen as disconnected, unexplored regions.

Disconnection is not seen in RVEA; RVEA's unique angle penalised distance (APD) scalarisation function gives it the ability to maintain uniformity across the population. Cheng [4] et al. showed that RVEA produces better quality Pareto front approximations than NSGA-III in multi-modal MOO problems, as it does in our results.

5   Case Study

The ultimate goal of experimental design is to create effective metamodels; therefore, the quality of metamodels should be verified as a means of determining optimisation success. We used our designs to explore the landscape of the ratio between time and molecular weight produced in the batch creation of branched polymers. Parameters for this simulation include $Time$, the duration of each batch production, M, monomer concentration, I, initiator, and T the temperature of the batch production vessel. For more details about the problem, see [17]. Bounds for each parameter can be seen in Table 1.

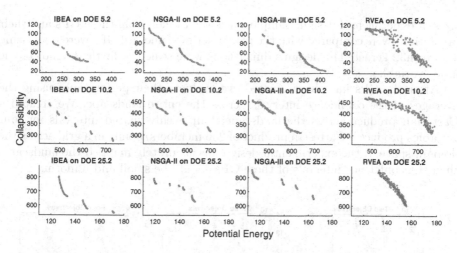

Fig. 3. Final populations for each algorithm, PlatEMO, two objective problems.

Table 1. Branched polymer input specifications.

Parameter	Range	Unit
Time	30 to 10,000	seconds
Monomer concentration	10 to 14	Meters cubed per second
Initiator	3E−5 to 1.5E−4	Meters cubed per second
Temperature	60 to 80	degrees centigrade

5.1 Multi-criteria Decision Making

Once optimisation is complete the DM can select a representative from the approximated Pareto set. In order to validate the success of the optimisation we select a design from the final population to use as an experimental design in the polymerisation problem. We used decomposition to select a choice. The weights for the four objectives are $[0.1, 0.1, 0.7, 0.1]$ (potential, discrepancy, collapsibility, correlation). For three objectives, $[0.3, 0.1, 0.6]$ (potential, discrepancy, collapsibility). In two objectives, $[0.4, 0.6]$ (potential and collapsibility). We considered collapsibility to be a very important property when exploring the landscape therefore a high weight was given. The weights suffer the disadvantage of human error, we cannot see all possibilities and must make assumptions. Investigations with other weights are not within the scope of this paper.

5.2 Results

The experiment was carried out with varying numbers of objectives on the four different algorithms. A Gaussian process (GP) was chosen to explore the outputs of the function because it is non-parameterised and the confidence intervals provide a good performance measure. A GP was fitted according to the various sizes of experimental design. Samples were then taken from these GPs and their predictions were compared with the true function value. All GPs were built using a matern32 kernel. The design's dimensions were scaled to fit the bounds of the input variables.

After the GPs have been created, we evaluate their performance using the averages of the confidence intervals across the entire landscape. We created a Cartesian product across the landscape; four evenly spaced intervals for four variables produce a Cartesian product of 256 members/points in a grid across the domain space. If the experimental design has accurately mapped the landscape, then the confidence intervals of these GPs should be small and uniform.

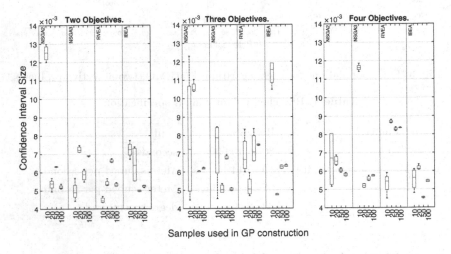

Fig. 4. Confidence intervals for the Cartesian evaluation, PlatEMO.

Figure 4 shows box-plots of the confidence intervals using the data from PlatEMO. No algorithm stands out as superior; however, the interquartile ranges (IQR) generally fall as the number of samples increases. In four objectives, many of the IQRs are low and thin. This suggests that the inclusion of more objectives produces more desirable qualities; treating DOE optimisation as a many-objective optimisation problem will produce better metamodels. In four objectives a correlation criterion is included, despite it not conflicting with other objectives the quality of the designs seems to improve. Suggesting that adding objectives that define good qualities, but from a different perspective, can add to the overall quality of the design. Three objectives has noticeably wider IQRs than two and four objectives. Work should be done to investigate whether it is the choice of criterions or the number of them that most effects performance; this result gives merit to both theories.

6   Conclusions

In this paper, we have explored the use of MOO as a tool for design of experiments. We successfully formulated the problem of DOE into a MOO problem. Designs have been successfully optimised and used to investigate a real-world problem where their success was verified. Various objective functions were chosen, potential energy, discrepancy, collapsibility, and correlation; these cover various desirable qualities. We have successfully investigated the performance of different MOEAs when optimising experimental designs. We chose four different algorithms to compare performance. Dominance-based approaches (NSGA-II), reference-based algorithms (RVEA and NSGA-III), and indicator-based algorithms (IBEA) were used. The high gene count and distance between initial and final population lead NSGA-II to converge better than NSGA-III in higher objectives. RVEA's unique scalarisation approach led it to perform well with high gene counts; IBEA performed well as a general use algorithm. We were successful in using optimised designs in the construction of metamodels. In the branch polymer metamodels, two and four objectives had low and narrow IQRs. Suggesting there is an optimal choice of criteria.

To further improve the knowledge of how best to utilise MOO in experimental design, further experiments should be conducted. Algorithms designed to tackle problems with large gene counts should be explored. Exploration of the use of more objectives/different objectives is needed to confirm how the choice of objectives effects the performance. More study into the performance of these designs in real life problems should be done. Ishibuchi et al. [12] discussed how for many-objective problems the number of solutions needed to best approximate the Pareto set becomes exponentially large; a bigger population is needed. We chose 200 individuals for our work however exploration of optimisations with higher population could be explored. Working on using different reproduction operators is also one of the future works. It is important to consider decision-maker's preferences before or after the optimisation process. This work finds a set of nondominated solutions and does not utilise preferences. Therefore, getting one solution based on the preferences (e.g., weights, desirable objective function values) will be in our future research.

References

1. Abdellatif, A.S., Abdelhalim, M.B., El Rouby, A.B., Khalil, A.H.: Hybrid Latin hypercube designs. In: 2010 The 7th International Conference on Informatics and Systems (INFOS), pp. 1–5 (2010)
2. Audze, P., Eglājs, V.: New approach to the design of multifactor experiments, problems of dynamics and strengths 35, pp. 104–107. Zinatne Publishing House (1977)
3. Bates, S., Sienz, J., Langley, D.: Formulation of the Audze-Eglais uniform Latin hypercube design of experiments. Adv. Eng. Softw. 34(8), 493–506 (2003)
4. Cheng, R., Jin, Y., Olhofer, M., Sendhoff, B.: A reference vector guided evolutionary algorithm for many-objective optimization. IEEE Trans. Evol. Comput. 20(5), 773–791 (2016)

5. Cheng, R., Jin, Y., Olhofer, M., Sendhoff, B.: Test problems for large-scale multi-objective and many-objective optimization. IEEE Trans. Cybernet. **47**(12), 4108–4121 (2017)
6. Deb, K.: Multi-Objective Optimization using Evolutionary Algorithms. Wiley, Chichester (2001)
7. Deb, K., Jain, H.: An evolutionary many-objective optimization algorithm using reference-point-based nondominated sorting approach, Part I: Solving problems with box constraints. IEEE Trans. Evol. Comput. **18**(4), 577–601 (2014)
8. Garud, S., Karimi, I., Kraft, M.: Design of computer experiments: a review. Comput. Chem. Eng. **106**, 71–95 (2017)
9. Gunpinar, E., Khan, S.: A multi-criteria based selection method using non-dominated sorting for genetic algorithm based design. Optim. Eng. **21**, 1319–1357 (2020)
10. Hickernell, F.J.: A generalized discrepancy and quadrature error bound. Math. Comput. **67**(221), 299–322 (1998)
11. Ishibuchi, H., Imada, R., Setoguchi, Y., Nojima, Y.: Performance comparison of NSGA-II and NSGA-III on various many-objective test problems. In: 2016 IEEE Congress on Evolutionary Computation (CEC), pp. 3045–3052 (2016)
12. Ishibuchi, H., Tsukamoto, N., Nojima, Y.: Evolutionary many-objective optimiza-tion: a short review. In: IEEE World Congress on Computational Intelligence, pp. 2419–2426 (2008)
13. Johnson, M., Moore, L., Ylvisaker, D.: Minimax and maximin distance designs. J. Stat. Plan. Infer. **26**(2), 131–148 (1990)
14. Joseph, V.R., Hung, Y.: Orthogonal-maximin Latin hypercube designs. Stat. Sin. **18**(1), 171–186 (2008)
15. Li, Y., Li, N., Gong, G., Yan, J.: A novel design of experiment algorithm using improved evolutionary multi-objective optimization strategy. Eng. Appl. Artif. Intell. **102**, 104283 (2021)
16. Miettinen, K.: Nonlinear Multiobjective Optimization. Kluwer, Boston (1999)
17. Mogilicharla, A., Chugh, T., Majumder, S., Mitra, K.: Multi-objective optimization of bulk vinyl acetate polymerization with branching. Mater. Manuf. Processes **29**, 210–217 (2014)
18. Peng, Y., Ishibuchi, H., Shang, K.: Multi-modal multi-objective optimization: prob-lem analysis and case studies. In: 2019 IEEE Symposium Series on Computational Intelligence (SSCI), pp. 1865–1872 (2019)
19. Zille, H., Mostaghim, S.: Comparison study of large-scale optimisation techniques on the LSMOP benchmark functions. In: 2017 IEEE Symposium Series on Com-putational Intelligence (SSCI), pp. 1–8 (2017)
20. Zitzler, E., Künzli, S.: Indicator-based selection in multiobjective search. In: Yao, X., et al. (eds.) PPSN 2004. LNCS, vol. 3242, pp. 832–842. Springer, Heidelberg (2004). https://doi.org/10.1007/978-3-540-30217-9_84

Evolutionary Continuous Optimization of Hybrid Gene Regulatory Networks

Romain Michelucci[✉][ID], Jean-Paul Comet[ID], and Denis Pallez[ID]

Université Côte d'Azur, CNRS, I3S, Sophia Antipolis, France
{romain.michelucci,jean-Paul.comet,denis.pallez}@univ-cotedazur.fr

Abstract. The study of gene regulatory networks (GRNs) allows us to better understand biological systems such as the adaptation of the organism to a disturbance in the environment. Hybrid GRNs (hGRNs) are of interest because they integrate the continuous time evolution in GRN modeling which is convenient in biology. This study focuses on the problem of identifying the variables of hGRN models. In a large-scale case, previous work using constraint-based programming has failed to solve the minimal constraints on such variables which reflect the biological knowledge on the system behavior. In this work, we propose to transform a Constraint Satisfaction Problem (CSP) into a Free Optimization Problem (FOP) by formulating an adequate fitness function and validate the approach on an abstract model of the circadian cycle. We compare several continuous optimization algorithms and show that these first experimental results are in agreement with the specifications coming from biological expertise: evolutionary algorithms are able to identify a solution equivalent to the ones found by continuous constraint solvers.

Keywords: Continuous single-objective optimization · Fitness formulation · hybrid GRN · Real-world application · Bio-inspired computation

1 Introduction

Genetic regulatory network (GRN) modeling aims at studying and understanding the molecular mechanisms that enable the organism to perform essential functions ranging from metabolism to environmental disturbance adaptation. Two types of control rules coexist in these regulatory networks: activations and inhibitions. Their combination allows the system to behave in a large variety of ways and the complexity of these systems comes from the so-called positive and negative feedbacks commonly observed, which respectively lead to multistationarity and homeostasis (ability to maintain a balance). Studying the dynamics of these systems opens new perspectives with crucial applications in fundamental biology, pharmacology, medicine, or chronotherapy for instance, which tries to choose the best time of day to administer the medication in order to limit the side effects while preserving the therapeutic effects.

P. Legrand et al. (Eds.): EA 2022, LNCS 14091, pp. 159–172, 2023.
https://doi.org/10.1007/978-3-031-42616-2_12

Numerous modeling frameworks have been proposed for representing GRNs such as differential frameworks (using ordinary differential equations), stochastic ones (considering that transitions between states have a stochastic nature), or discrete ones (modeling the presence or absence of biological entities in the system states). Even if each of them presents their own advantages, they all rely on the identification of the variables that govern the model dynamics and this variable identification remains the limiting step. To address this difficulty, a considerable number of research groups apply evolutionary algorithms to fit GRN models and variables to gene expression data, see e.g. the survey [16].

In the present work, we prefer to consider *hybrid* frameworks [1], called hGRNs, which add to discrete ones [17] the time spent in each of the discrete states. Once more, the variables' identification remains the bottleneck of the modeling process, but one can seek in such a hybrid framework for an automation of this step to build a model in agreement with the experimental observations. Indeed, modeling variations of protein concentration in a biological system can be very hard for numerous proteins. Nevertheless, experimental observations allow us to represent experimental traces by irregularly spaced time series of observable events. From those events, minimal constraints on the hGRN variables can be deduced and the authors attempted to use continuous Constraint Satisfaction Problem (CSP) solvers [2] but faced difficulties in extracting solutions.

In this paper, we show that the constraint problem, which characterizes the set of solutions exhaustively, can be expressed as a FOP [6,8] by indirectly handling constraints. More precisely, the representation of biological knowledge as a sequence of observable events allows to define a high-dimensional non-trivial continuous optimization problem in which the search space increases exponentially with the number of genes involved in the hGRN.

The work focuses on the FOP formulation, on the fitness characterization and performs some comparisons between several bio-inspired algorithms, leaving out the scalability problem which is out of the scope of the article. We illustrate the approach on a very abstract model of the circadian cycle (subsystem allowing an adaptation of the body to day/night alternation).

The paper is organized as follows: Sect. 2 describes the models used for representing the dynamics of biological systems and the biological knowledge used as an input. Section 3 proposes a method whereby the modeling problem is reformulated as a continuous FOP that can be solved by means of a bio-inspired algorithm. Experimental results are discussed in Sect. 4 and some conclusions are drawn in Sect. 5.

2   Problem Description

2.1   Hybrid GRN

To build a digital model of a biological system, it is necessary to know precisely how it works. Such a system is defined as a set of genes performing a biological function and represented in the form of a GRN where vertices V correspond to an abstraction of one or more biological genes (within circles) and edges

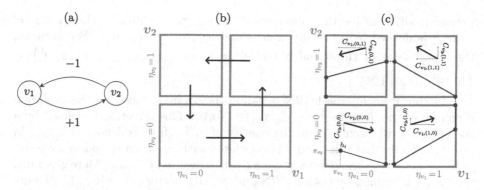

Fig. 1. Interaction graph representing the circadian cycle (a), its discrete state graph (b) and a possible dynamic of its hybrid state graph (c).

depicting activations ($+$) or inhibitions ($-$). It can be statically represented as a labeled directed graph or *interaction graph* (cf. Fig. 1a). For studying the GRN evolution, we first need to define the system state as the concentrations vector of the proteins related to genes. Because the regulations take place above particular thresholds, we associate with the sign of the regulation an abstract threshold: $v_1 \xrightarrow{+n} v_2$ (resp. $v_1 \xrightarrow{-n} v_2$) means that v_1 can activate (resp. inhibit) v_2 only if the concentration of v_1 is above its nth threshold (ranked by increasing order). For example, graph of Fig. 1a forms a negative feedback loop where each gene ($v_1, v_2 \in V$) has an indirect negative action on itself: when v_1 is active, it is above its first threshold (we note $v_1 = 1$), then, v_1 activates the gene v_2 and v_2 passes from level 0 (under its first threshold) to another level greater than its first threshold ($v_2 = 1$). As v_2 reaches level 1, v_2 inhibits v_1, and so on. This represents a highly abstracted model of regulations piloting the circadian cycle ensuring the cyclic adaptation (day or night) of the organism.

In order to integrate dynamics in the previous model, the first step is to enumerate all possible states: a *discrete state* is defined by the level of all genes contained in the GRN. Thus, if there are n genes, each state η is defined by a vector of n integers ($\eta_{v_1}, ..., \eta_{v_n}$) and $\mathbb{S}$ denotes the set of all possible discrete states of the GRN. For instance, state $(0, 0)$, the bottom left gray square box in Fig. 1b, corresponds to the state where discrete levels of genes v_1 and v_2 are both equal to 0. The second step consists in adding transitions between all these states (black arrows). Thus, *state graph* of Fig. 1b represents the dynamics associated with the interaction graph of Fig. 1a. Such kind of models is very interesting for logically reasoning on regulatory changes. Nevertheless, this qualitative modeling framework totally abstracts time information whereas, for numerous biological systems, time plays a crucial role in the system's fate.

In addition to discrete transitions (dotted lines between points in Fig. 1b), an hGRN adds continuous evolution of gene product concentration in each discrete state represented by a continuous trajectory (see piecewise linear solid lines). One point on this trajectory inside a particular discrete state, is given by a precise position inside the square: $\pi = (\pi_{v_1}, ..., \pi_{v_n}) \in [0, 1]^n$. Thus, a *hybrid state* h is defined by a discrete state η and its *fractional part* π. For instance, the coordinates of the initial hybrid state h_i are $\left((\eta_{v_1}, \eta_{v_2})^t, (\pi_{v_1}, \pi_{v_2})^t\right) = \left((0, 0)^t, (0.25, 0.25)^t\right)$.

Starting from h_i, the hGRN dynamics is given by following the evolution direction of the discrete state $(\eta_{v_1}, \eta_{v_2}) = (0, 0)$. This direction is defined by a so-called *celerity vector*. Thus, the celerity of v_1 in $(0, 0)$ is denoted $C_{v_1, (0,0)}$ in order to specify that this celerity is associated with v_1, when v_1 and v_2 levels are 0. In a similar way, the celerity of v_2 in $(0, 0)$ is denoted $C_{v_2, (0,0)}$. More generally, an hGRN is defined by both a GRN and celerity vectors $C = \{C_{v,\eta}\}$, a family of floated values indexed by (v, η) where $v \in V$ and $\eta \in \mathbb{S}$. $C_{v,\eta}$ is called the *celerity* of v in η. The hybrid state graph of Fig. 1b depicts one possible dynamic associated with the interaction graph of Fig. 1a. Starting from the initial hybrid state h_i, v_1 concentration increases until it reaches the right border of discrete state $(0, 0)$. From this border, the trajectory jumps into the neighbor state $(1, 0)$ because the celerity vector of this second state does not oppose the entry of the trajectory (signs of v_1 celerities in both states are the same). In $(1, 0)$, the trajectory reaches the right border of this discrete state which corresponds to the maximum admissible concentration of v_1. As there is no discrete state at the right of $(1, 0)$, the trajectory evolves on this border in v_2 direction resulting in a so-called *slide* of v_1, noted $slide^+(v_1)$. After sliding, the trajectory enters the state $(1, 1)$. This process follows up until the trajectory enters back the initial state $(0, 0)$. The complete definition of hGRN dynamics can be found in [1].

Such modeling frameworks are very useful to reason on the GRN trajectories. Nevertheless, as usual, the bottleneck of the modeling process relies on the determination of variable values controlling the trajectories, that is the celerities. The goal of this paper is to automatically determine, from some formalized biological information, all celerity vector values in order to obtain a valid hGRN model of the biological system studied. In the next part, we introduce the *biological knowledge* (BK) from which celerity values can be determined.

2.2 Biological Knowledge

As opposed to numerous works that attempt to automatically build a model from raw experimental data [3,5,12,15], the present work takes into consideration already-formalized information analyzed by biologists themselves coming from both biological data and expertise. This complementary approach is preferred because raw data are subject to noisiness and scarcity. A biological experiment consists of (i) putting the biological system in a particular initial state partially defined, (ii) recording the sequence of observable events, and (iii) measuring the reached final state of the observed system. While initial and final states are described using their discrete and fractional parts $h_i = (\eta_i, \pi_i)$ and $h_f = (\eta_f, \pi_f)$, a sequence of observable events is formalized by a sequence of triplets of the form $(\Delta t, b, e)$. Each element of each triplet expresses a property on the behavior in the current discrete state: Δt delineates the time spent in the current state; b

specifies the observed behaviors during the continuous trajectory expressed by $slide(v)$ and $noslide(v)$; finally, e represents the next discrete state transition which is of the form $v+$ (resp. $v-$) specifying that the next discrete event is the increasing (resp. decreasing) of the discrete level of v.

For the interaction graph of Fig. 1a, biological expertise can be summarized as follows: there exists a behavior starting from a particular point of coordinates going through four discrete states and coming back to the initial point after 24 h. More precisely, the time spent in each of the 4 discrete states is approximately 5 h in $(0,0)$, 7 in $(1,0)$, and so on. See the first properties of each event in the following description of the biological knowledge:

$$\{h_i\} \begin{pmatrix} 5.0 \\ noslide\,(v_2) \\ v_1+ \end{pmatrix} ; \begin{pmatrix} 7.0 \\ slide^+\,(v_1) \\ v_2+ \end{pmatrix} ; \begin{pmatrix} 8.0 \\ noslide\,(v_2) \\ v_1- \end{pmatrix} ; \begin{pmatrix} 4.0 \\ slide^-\,(v_1) \\ v_2- \end{pmatrix} \{h_f\} \quad (1)$$

where $h_i = ((0,0)^t, (0.0, 1.0)^t)$ is the initial hybrid state and h_f (final hybrid state) is equal to h_i. For the first event, v_1+ constrains the trajectory to reach the next discrete state by increasing the concentration level of v_1. The second property $noslide(v_2)$ in $(0,0)$ expresses that the trajectory has to reach the right border of the discrete state without touching the upper or lower borders as explained in Sect. 2.1. The continuous trajectory of Fig. 1b satisfies all properties of eq. (1) except for the initial point h_i which is misplaced: it should be located in the top left corner of discrete state $(0,0)$ to allow trajectory to be a cycle.

Figure 2 represents for each discrete state, and one after another all possible trajectories satisfying eq. (1) using shaded surfaces. Starting from h_i, the surface represents all compatible celerity vectors of $(0,0)$ which lead the trajectory to the next expected state without sliding at the bottom or top border. For illustrative purposes, two instances of compatible trajectories are highlighted: a solid one and a dotted one.

2.3   Constraint Satisfaction Problem (CSP) Approach

Our goal is to identify celerity vectors that define trajectories (cf. Sect. 2.1) satisfying constraints given by the *biological knowledge BK*. An earlier attempt has been developed using constraint-based programming [2]. This CSP formulation led to constraints on celerity vectors which had to be satisfied for the hGRN dynamics to be consistent with BK. However, the exploitation of the constraints generated was not so easy: classical solvers were not able to extract particular solutions. Let us consider a CSP that aims to find all solutions satisfying the constraint $y \leq x^2$. A continuous solver paves the search space in multiple tiles (shaded rectangles in Fig. 3). Light tiles only contain solutions of the CSP whereas dark tiles may contain values that do not satisfy the constraint (i.e. $y > x^2$ above the curve).

The problem that arises from using a continuous solver may be summed up by its inability to extract particular solutions on the function curve. It would be necessary to obtain a tiling of infinitesimal size. That is why we decided to reformulate the hGRN variables' identification as an optimization problem.

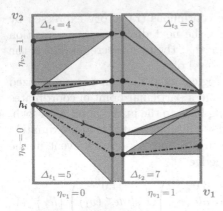

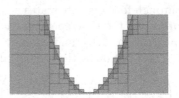

Fig. 2. Visual representation of the infinite set of possible solutions.

Fig. 3. CSP difficulty to target solutions for constraint $y \leq x^2$.

2.4   Problem Characterization

Finding celerity values that satisfy BK constraints consists of finding a continuous trajectory that (i) goes through the right sequence of discrete states, (ii) spends the right elapsed time in each encountered state, and (iii) satisfies the right behavior in each state by sliding or not. In the case of a trajectory that does not satisfy BK, we measure how much it does not respect this knowledge. For instance, as BK specifies spending 5 h in $(0,0)$, a trajectory spending 5 h and 10 min is "better" than a trajectory that only spends 2 h in the same discrete state. In other words, we use the notion of *distance* between a trajectory and the expected properties expressed by BK: this distance vanishes as soon as all properties of BK are satisfied. Since BK specifies the properties of a sequence of states, we can decompose such distance by computing how a considered trajectory tr inside each state η is far from BK properties of the corresponding state. Thus, the global distance of one property p is defined by summing such distances $d_{p,\eta}$ inside each encountered discrete state $\eta \in \mathbb{S}$ where p is one of the three BK properties Δt, b, or e. Therefore, we define three criteria:

Time criterion. The first criterion $d_{\Delta t}$ is related to the time spent in the current discrete state. It is the Euclidean distance between the expected time t_η^* of BK and the time t_η necessary for the current trajectory to reach the exit point from the current state:

$$d_{\Delta t}(tr, BK) = \sum_{\eta} d_{Euclidean}\left(t_\eta, t_\eta^*\right) \qquad (2)$$

Slide criterion. Second criterion evaluates the distance between the continuous trajectory behaviors inside each encountered discrete state and the properties of sliding in BK (denoted b in each observable event). Three different cases are considered and respectively illustrated in Fig. 4 where shaded surfaces and dotted lines represent BK and black dotted double arrows, the distance d_b: (i)

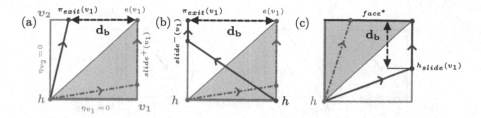

Fig. 4. Illustration of evaluation cases with respect to BK behavior property.

"v should slide according to BK, but the trajectory tr does not". In this case (Fig. 4a), we compute the difference between the fractional part of the exit point of v according to tr ($\pi_{exit}(v)$) and $e(v)$ which is the fractional part of the exit point according to the sliding BK property (it either equals to 0 when $slide^-(v)$ or 1 when $slide^+(v)$):

$$d_{b,\eta}(tr, BK) = |\pi_{exit}(v) - e(v)| \tag{3}$$

where v is the gene concerned by the sliding property of the current discrete state η; (ii) "v should slide on max (resp. min) level according to BK, but the given trajectory slides on min (resp. max)". We consider it (see Fig. 4b) as a special case of previous item (eq. (3)) where the exit point of the trajectory $\pi_{exit}(v)$ is either equal to 0 (sliding right in Fig. 4b) or 1 (sliding left in Fig. 4b); (iii) "v should not slide according to BK, but tr does" (Fig. 4c). Here we compute the Manhattan distance between the first hybrid state where v begins to slide $h_{slide}(v)$ and the expected exit face noted $face^*$:

$$d_{b,\eta}(tr, BK) = d_{Manhattan}(h_{slide}(v), face^*) \tag{4}$$

In Fig. 4c, the expected exit face is the north one (bold line). As for the previous criterion, $d_b(tr, BK)$ is defined as the sum of the different $d_{b,\eta}(tr, BK)$ for each encountered discrete state η.

Discrete criterium. Intuitively, we have to compare the expected next discrete state (according to BK) with the discrete one into which the given trajectory tr enters. Unfortunately, in some cases, it is not possible to compute tr next discrete state because the trajectory can be *blocked* in the current discrete state. Let us take as an example the situation where the celerity vector inside the $(0,0)$ discrete state points towards the south-west direction (cf. Fig. 5a). The trajectory is blocked because the concentration of both gene products vanishes and there are no neighbors in these directions. In order to accurately evaluate tr, following the sequence of discrete states of BK, we evaluate the local distance between the considered trajectory inside the current discrete state and the associated BK. If the given trajectory does not allow the right discrete transition, then we artificially restart the trajectory in the next expected discrete state of BK.

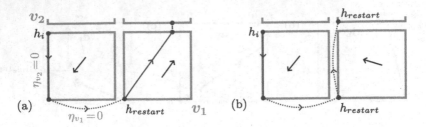

Fig. 5. Illustration of blockage (a) and wrong discrete transition (b).

The initial restart point $h_{restart}$ is defined by the new discrete state combined with the same fractional part before it stopped. This step is illustrated by the curved dotted lines (cf. Fig. 5). Therefore, d_e has to take into account, on the one hand, the Manhattan distance between the expected next discrete state η^{+^*} and the next discrete state η^+ according to tr and, on the other hand, the number of detected blockages:

$$d_e(tr, BK) = \sum_\eta \left(d_{Manhattan}\left(\eta^+, \eta^{+^*}\right) + 1\!\!1_{blockage}(\eta) \right) \tag{5}$$

where $blockage(\eta)$ is $True$ if the trajectory is blocked in the current discrete state η, and η^+ (resp. η^{+^*}) is the next discrete state according to tr (resp. to BK). Note that when a blockage occurs η^+ is not defined and in such a case $d_{Manhattan}\left(\eta^+, \eta^{+^*}\right)$ is considered zero (the penalty comes from $1\!\!1_{blockage}(\eta)$).

Aggregating Criteria. We are focusing on formulating an adequate fitness function by indirectly handling constraints. Constraints are embedded into the three previously described optimization criteria such that all we need to care about is optimizing them. Thus, identifying celerity values consists in minimizing these criteria. One could consider this problem as a multi-criteria optimization problem. However, they are neither conflicting nor invariant: solutions exist that simultaneously optimize each criterion. Therefore, we suggest to combine them into a global distance $g(tr, BK)$ which consists in a combination of $d_{\Delta t}(tr, BK)$, $d_b(tr, BK)$ and $d_e(tr, BK)$ where the criteria weights are equal. Minimizing g leads to a single-objective optimization problem and will be addressed using bio-inspired algorithms. We propose two versions of the aggregation of three different criteria: an additive version defined by $g_+ = d_{\Delta t} + d_b + d_e$ and a multiplicative version defined by $g_\times = (1 + d_{\Delta t}) \times (1 + d_b) \times (1 + d_e) - 1$. Although the former is commonly used, the latter is proposed because, intuitively, it could have a greater impact on the convergence rate: errors are amplified and improvements are better controlled thanks to a steeper gradient. As each distance should be as close to 0 as possible, g_+ (resp. $g_\times$) should also be as close to 0 as possible (resp. thanks to the subtraction of 1). That leads to the definition of two fitness functions (knowing that BK is fixed):

$$f_+(x) = g_+(tr, BK) \quad (6) \qquad\qquad f_\times(x) = g_\times(tr, BK) \quad (7)$$

whose domain is $\left(\prod_{v \in V}[0, b_v]\right) \times [0, 1]^n \times \mathbb{R}^{|C|}$ and codomain $\mathbb{R}^+$.

3   Bio-Inspired hGRN Modeling Search

This section presents different bio-inspired approaches for identifying celerities of an hGRN. For this purpose, we compare several continuous single-objective bio-inspired algorithms for searching trajectories that satisfy biological knowledge BK as explained in Sect. 2.4.

Representation. As presented in Sect. 2.2, a trajectory is characterized by all celerities of all discrete states $\{C_{v,\eta}\}$ plus the initial hybrid state h_i. Thus, trajectory genotype of Fig. 1b is defined by a tuple of 2 integers and 2 float values for h_i and 8 float values for celerities: the genotype is represented by $x = (h_i; C_{v_1,(0,0)}; C_{v_2,(0,0)}; C_{v_1,(1,0)}; C_{v_2,(1,0)}; C_{v_1,(1,1)}; C_{v_2,(1,1)}; C_{v_1,(0,1)}; C_{v_2,(0,1)})$.

Each floated value varies in the interval $[-r; r]$ with r equals 2 by default. In the presented example, the problem of identifying variables of an hGRN may seem trivial, nevertheless, in realistic models, the size of the genome is exponential with respect to the number n of genes: the initial hybrid state $h_i = (\eta_i, \pi_i)$ is described by n integer values for the discrete state and n float values for the fractional part. Because the number of celerities is also equal to n in each state and because the number of states is $|\mathbb{S}| = \prod_{v \in V}(b_v + 1)$, the total number of celerities $|C|$ is at most $n \times |\mathbb{S}| = n \times \prod_{v \in V}(b_v + 1)$ (possibly less in case of equality of a priori different celerities).

Fitness Evaluation. Evaluating a candidate solution consists in computing the difference between BK formalized in 2.2 and the given trajectory obtained from celerities contained in the genome. To do so, we simulate the trajectory thanks to the initial state h_i and evaluate, discrete state by discrete state, each of the three introduced criteria $d_{\Delta t}$, d_b, and d_e.

Continuous Optimization Methods. A baseline random optimization (RO) [11] and the following four continuous meta-heuristic algorithms are compared:

(i) Differential Evolution (DE) [14], a global search heuristic using a binomial crossover and a mutation operator of $DE/rand/1/bin$. The different control parameters are $P_{CR} = 0.3$ and F is selected from the interval $[0.5, 1.0]$ randomly for each difference vector with the dither technique.

(ii) a simple $(\mu + \lambda)$ Genetic Algorithm (GA), used with a binary tournament selection and the following operators: Simulated Binary Crossover and Polynomial Mutation are applied with Fitness Survival. All duplicates are removed.

(iii) Adaptive Particle Swarm Optimization (APSO) [18] which is based on the simulating of social behavior. The algorithm uses a swarm of particles to guide its search. Each particle has a velocity and is influenced by locally and globally best-found solutions. The default parameters are $w = 0.9, c_1 = 2.0, c_2 = 2.0$ with a $max\_velocity\_rate = 0.2$.

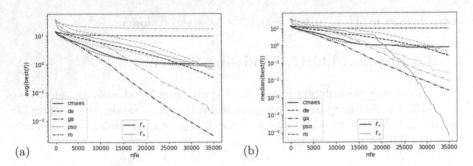

Fig. 6. Comparison of monotonic evolution of (a) mean and (b) median best fitness values by algorithm and fitness function on 100 runs. The y-axis is log scaled.

(iv) Covariance Matrix Adaptation Evolutionary Strategy (CMA-ES) [9], which is a state-of-the-art and self-adaptive EA with the default initial standard deviation in each coordinate $\sigma = 0.1$.

4 Experimental Study

The four meta-heuristics are implemented in pymoo [4]. To evaluate the algorithms' performance, we execute 100 independent runs for each algorithm and each fitness function. An initial population size of 500 is applied, followed by 35000 function evaluations (NFE). Both experiments are realized on the hGRN of Fig. 1b using BK described by eq. (1) with h_i fixed to $((0,0)^t, (0.0, 1.0)^t)$.

Results. For each algorithm and each fitness function, at each generation we compute the best candidate solution so far, repeat 100 times the executions and compute the mean (resp. median) over the 100 runs. Monotonic evolutions of all algorithms are depicted in Fig. 6 where dark lines represent f_+ and light lines, $f_\times$. It can be observed that (i) as expected, meta-heuristics results are (far) better performing compared to RO algorithm, (ii) decreases of f_+ and $f_\times$ values are done at the same pace (the curves are roughly parallel), except for CMA-ES whose $f_\times$ median evolution has a better convergence rate than with f_+, and (iii) apart from this case, GA convergence of the fitness function is one of the best (with both f_+ and $f_\times$) when focusing on the mean (resp. median).

Table 7a summaries statistics of the results obtained after 100 runs of the five considered algorithms. The best result (column by column) for f_+ (resp. $f_\times$) is bolded. Minimum, average and standard deviation are reported along with the Biological Success Rate (BSR) defined by the number of times an algorithm finds a solution with a fitness close to 0 with a precision error ϵ equal to 10^{-2}. BSR is based on the traditional success rate but introduces an important precision error coherent with biological expertise. For instance, a trajectory which would slide in $\eta = (0,0)$ during a fraction of seconds ($< \epsilon$) very next to the exit point $e(v_1) = 1.0$ before going to the next discrete state is an acceptable trajectory

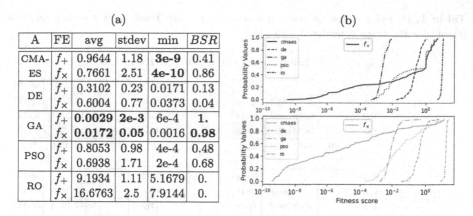

(a)

A	FE	avg	stdev	min	BSR
CMA-ES	f_+	0.9644	1.18	**3e-9**	0.41
	$f_\times$	0.7661	2.51	**4e-10**	0.86
DE	f_+	0.3102	0.23	0.0171	0.13
	$f_\times$	0.6004	0.77	0.0373	0.04
GA	f_+	**0.0029**	**2e-3**	6e-4	**1.**
	$f_\times$	**0.0172**	**0.05**	**0.0016**	**0.98**
PSO	f_+	0.8053	0.98	4e-4	0.48
	$f_\times$	0.6938	1.71	2e-4	0.68
RO	f_+	9.1934	1.11	5.1679	0.
	$f_\times$	16.6763	2.5	7.9144	0.

Fig. 7. Summary (a) and CDF curves (b) of overall best results.

despite BK stating $noslide(v)$. In addition, Cumulative Distribution Function (CDF) curves are constructed in Fig. 7b for f_+ (top) and $f_\times$ (bottom). Each CDF curve describes the probability that a solution is found at, or below, a given fitness score. For instance, in $f_\times$ experiment, there is almost 60% probability that a user obtains a solution at a fitness score less than or equal to 10^{-4} with CMA-ES (given 35000 NFE). From both diagrams, two algorithms stand out: GA has the highest probability to obtain good results and there is a non-zero probability for CMA-ES to perform top results ($< 10^{-8}$).

To statistically validate the observed differences among the algorithms, we conducted a statistical validation campaign on the reported performance values of the two following scenarios: (i) algorithms performances obtained with f_+ objective function and (ii) algorithms performances achieved with $f_\times$ one. In addition, a third scenario is suggested as being a comparison of algorithms performances between f_+ and $f_\times$. First, we employ the Friedman rank-sum test [10] to assess whether at least two algorithms exhibit significant differences in the observed performance values. The p-values for the null hypothesis are $p_+ = 5e-56$ and $p_\times = 2e-64$ for f_+ and $f_\times$ respectively. At the 0.05 confidence level, the differences among the algorithms are significant. The statistical analysis proceeds with a post hoc analysis to determine which pairs of algorithms show significant differences in performance (for the three scenarios considered). In this step, we proceed to the Wilcoxon signed-rank test (as neither normality nor homoscedasticity conditions required for the application of parametric tests hold [7]) on the performance samples of each pair of algorithms. In addition, to reduce the issue of having Type I errors given multiple comparisons, the Bonferroni correction method is applied.

For all scenarios, Table 1 present tile-plots to illustrate all pairwise differences in the observed performance samples at the 0.05 confidence level. More specifically, the outcomes of the pairwise Wilcoxon-signed rank tests, without and with the application of the Bonferroni correction method, are provided on the left and right-hand side of the table respectively. Each tile corresponds to a pairwise sig-

Table 1. Pairwise Wilcoxon statistical tests (left) with Bonferroni post hoc analysis (right) for the three considered scenarios.

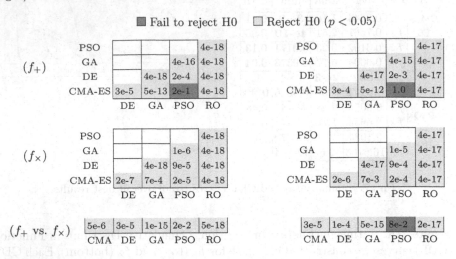

■ Fail to reject H0   □ Reject H0 ($p < 0.05$)

(f_+) — left:

	DE	GA	PSO	RO
PSO				4e-18
GA			4e-16	4e-18
DE		4e-18	2e-4	4e-18
CMA-ES	3e-5	5e-13	2e-1	4e-18

(f_+) — right:

	DE	GA	PSO	RO
PSO				4e-17
GA			4e-15	4e-17
DE		4e-17	2e-3	4e-17
CMA-ES	3e-4	5e-12	1.0	4e-17

$(f_\times)$ — left:

	DE	GA	PSO	RO
PSO				4e-18
GA			1e-6	4e-18
DE		4e-18	9e-5	4e-18
CMA-ES	2e-7	7e-4	2e-5	4e-18

$(f_\times)$ — right:

	DE	GA	PSO	RO
PSO				4e-17
GA			1e-5	4e-17
DE		4e-17	9e-4	4e-17
CMA-ES	2e-6	7e-3	2e-4	4e-17

$(f_+ \text{ vs. } f_\times)$ — left:

CMA	DE	GA	PSO	RO
5e-6	3e-5	1e-15	2e-2	5e-18

$(f_+ \text{ vs. } f_\times)$ — right:

CMA	DE	GA	PSO	RO
3e-5	1e-4	5e-15	8e-2	2e-17

nificance test between the algorithms of the corresponding row and column. The color of the tile indicates if the observed performance differences were enough to reject the null hypothesis at the significance level (p-value < 0.05). Light gray tiles indicate significant differences between the pair of algorithms, while dark gray tiles indicate that no significant differences were observed. Analyzing these results, if we base acceptance or rejection of the above hypotheses, we arrive at the following insights: (i) in f_+ scenario PSO performances are not significantly different and (ii) Bonferroni correction reveals that PSO performances are the same whatever fitness function. Nevertheless, the performances of other algorithms depend on the chosen fitness function. Therefore, according to the algorithm considered, the fitness function choice has definitely an impact on the performances: f_+ is preferred when considering DE and GA while $f_\times$ is in the case of CMA-ES and PSO.

Finally, with respect to the conducted experiments, GA and CMA-ES will be investigated in the future as the first one gives good and stable results with high probability, whereas the second performs better overall (the best solutions are obtained using CMA-ES), but is subject to instability (due to exploration phases).

Visualization of the results. The application of bio-inspired algorithms allows us to exhibit different solutions consistent with BK and they seem complementary to the CSP approach. Both diagrams of Fig. 8 present as solid lines the overall best trajectory obtained by GA (a) and CMA-ES (b) together with the dotted one, obtained by the CSP approach using the CSP solver Absolute [13] combined with a possible strategy for cutting the search space [2]. The solutions provided by GA and CMA-ES illustrate the diversity of acceptable solutions

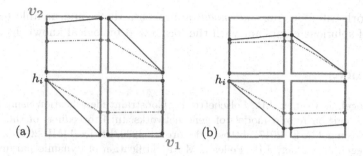

Fig. 8. Best trajectories (solid) obtained by GA (a) and CMA-ES (b) with f_+ compared to one of the solutions obtained by the CSP approach (dotted).

that are compliant (the structure of the trajectories is similar) with BK. From a modelization perspective, it would be great to exhibit a diverse sampling of possible solutions, in order to reason not only on one possible identification but on a set of sensible identifications.

5   Conclusion

The goal of this paper is to show that the problem of identifying variables in an hGRN, already formalized as a CSP, can be transformed into a bio-inspired optimization problem.

In previous works, many biological experiments have been formalized as constraints on time, behavior, and discrete events with the help of biologists' expertise. From these constraints, our work focused on finding how to model them as an FOP: we proposed a representation of a candidate solution and designed two appropriate fitness evaluation functions. To empirically test our approach, we conducted a study with a random optimization algorithm and four well-known continuous meta-heuristics: the proposed method shows satisfying results as the newly introduced BSR metric is high. In our experiments, CMA-ES obtains the overall best solutions satisfying BK constraints. Nevertheless, for this kind of problem, GA appears to be the best meta-heuristic because of its high probability of getting good results.

The proof-of-concept developed in this paper will shortly be applied to designing a new cell cycle hGRN model where time plays a crucial role in passing through each phase. Although this cell cycle model contains only 5 abstract genes, the number of celerities is about 240. The optimization problem will be challenging and lead us to apply *large-scale* optimization algorithms.

Moreover, when working with biologists, our ability to propose different solutions compliant with BK is of great importance because it leads to considerate new information which would not exhibit otherwise. Diversity in solutions reflects, on the one hand, a plurality of functioning within an observed system and, on the other hand, helps to evaluate the robustness of oscillating biological systems (the more diversity, the more robustness). From such a perspective, future work will focus on *multimodal* approaches that could be able to sample the set of solutions compliant with the formalized biological knowledge.

References

1. Behaegel, J., Comet, J.P., Folschette, F.: Constraint identification using modified Hoare logic on hybrid models of gene networks. In: Proceedings of the 24th Int. Symposium TIME (2017). https://doi.org/10.4230/LIPIcs.TIME.2017.5
2. Behaegel, J., Comet, J.P., Pelleau, M.: Identification of dynamic parameters for gene networks. In: Proceedings of the 30th IEEE International Conference on ICTAI (2018)
3. Biswas, S., Acharyya, S.: Neural model of gene regulatory network: a survey on supportive meta-heuristics. Theory Biosci. (2016). https://doi.org/10.1007/s12064-016-0224-z
4. Blank, J., Deb, K.: Pymoo: multi-objective optimization in python. IEEE Access. **8**, 89497–89509 (2020)
5. Buchet, S., Carbone, F., Magnin, M., Ménager, M., Roux, O.: Inference of Gene Networks from Single Cell Data through Quantified Inductive Logic Programming (2021). https://doi.org/10.1145/3486713.3486746
6. Coello, C.A.C.: Constraint-handling techniques used with evolutionary algorithms. In: Proceedings of GECCO (2021). https://doi.org/10.1145/3449726.3461400
7. Eftimov, T., Korošec, P.: Statistical Analyses for Meta-Heuristic Stochastic Optimization Algorithms: GECCO Tutorial (2020). https://doi.org/10.1145/3377929.3389881
8. Eiben, A.E., Smith, J.E.: Constraint Handling (2015). https://doi.org/10.1007/978-3-662-44874-8_13
9. Hansen, N.: The CMA Evolution Strategy: A Comparing Review (2006). https://doi.org/10.1007/3-540-32494-1_4
10. Hollander, M., Wolfe, D.A., Chicken, E.: Nonparametric statistical methods (2013)
11. Matyas, J., et al.: Random optimization. Automation and Remote control (1965)
12. Mitra, S., Biswas, S., Acharyya, S.: Application of meta-heuristics on reconstructing gene regulatory network: a Bayesian model approach. IETE J. Res. (2021). https://doi.org/10.1080/03772063.2021.1946433
13. Pelleau, M., Miné, A., Truchet, C., Benhamou, F.: A constraint solver based on abstract domains. In: 14th International Conference on VMCAI 2013 (2013). https://doi.org/10.1007/978-3-642-35873-9_26
14. Price, K., Storn, R.M., Lampinen, J.A.: Differential Evolution: A Practical Approach to Global Optimization (Natural Computing Series) (2005)
15. da Silva, J.E.H., Betnardino, H.S., Helio J.C., B., Vieira, A.B., Luciana C.D., C., de Oliveira, I.L.: Inferring gene regulatory network models from time-series data using metaheuristics. In: IEEE CEC (2020). https://doi.org/10.1109/CEC48606.2020.9185572
16. Spirov, A., Holloway, D.: Using evolutionary computations to understand the design and evolution of gene and cell regulatory networks. Methods (2013). https://doi.org/10.1016/j.ymeth.2013.05.013
17. Thomas, R.: Boolean formalization of genetic control circuits. J.T.B. (1973)
18. Zhan, Z.H., Zhang, J., Li, Y., Chung, H.S.H.: Adaptive particle swarm optimization. IEEE Trans. Syst. Man Cybern. Part B (Cybern.) **39**, 1362–1381 (2009). https://doi.org/10.1109/TSMCB.2009.2015956

Designing Attention Based Convolutional Neural Network (CNN) Architectures for Medical Image Classification Using Genetic Algorithm Based on Variable Length-Encoding Scheme

Muhammad Junaid Ali$^{(\boxtimes)}$, Laurent Moalic, Mokhtar Essaid, and Lhassane Idoumghar

Université de Haute-Alsace, IRIMAS UR 7499, 68093 Mulhouse, France
muhammad-junaid.ali@uha.fr,
{laurent.moalic,mokhtar.essaid,lhassane.idoumghar}@uha.fr

Abstract. Automatic diagnosis of abnormalities and diseases using medical scans consisting of different modalities (X-rays, mammograms, Optical Coherence Tomography (OCT)) is a challenging task due to changing clinical environment and varying noise levels. Manually designing deep learning architectures is a tedious task. However, Neural Architecture Search (NAS) provides the flexibility to automatically search for a suitable architecture for a given problem. In this paper GAMED-A-CNN, a Genetic Algorithm (GA)-based NAS approach is proposed for the medical image classification problem. The proposed algorithm is applied on different datasets considering multiple performance measures, where the effectiveness of the proposed approach was demonstrated. Furthermore, a variable-length encoding scheme is used for the representation of CNN architecture. The convolution attention layer is also used in the search space, which focuses on salient regions in the images to improve the classification performance. The comparison shows that the proposed approach achieves equal or superior performance compared to the best-known approaches.

Keywords: medical image classification · neural architecture search · genetic algorithm · automatic machine learning · deep learning · visual attention

1 Introduction

Advances in Convolutional Neural Networks (CNNs) enable research on more complex and advanced computer vision topics such as object detection, classification and segmentation [1]. CNN has achieved state-of-the-art performance on medical image diagnosis tasks, where the nature of data is complex [2]. Various CNN architectures have been proposed in recent years for different application domains.

Designing CNN architectures using a hit-and-trial approach is a tedious task. because the researchers needs to try combinations of different layers to create

© The Author(s), under exclusive license to Springer Nature Switzerland AG 2023
P. Legrand et al. (Eds.): EA 2022, LNCS 14091, pp. 173–186, 2023.
https://doi.org/10.1007/978-3-031-42616-2_13

these architectures, which requires a lot of time. The famous ResNet-50 architecture that uses residual connections to overcome the overfitting problem consists of 50 layers and has over 23 million trainable parameters [15]. Likewise, EfficientNet-b0 architecture consists of 5.3 million parameters. These architectures are trained on large-scale image recognition datasets with multiple classes, which can also be used as a Transfer Learning (TL) task for medical image analysis. In TL, the pre-trained model is fine-tuned into some different relevant tasks.

However, training these architectures on medical datasets having different characteristics does not guarantee optimal performance. Specific architectures are designed to achieve better performance on these tasks having smaller architecture sizes and fewer parameters. The Neural Architecture Search (NAS) approach consists of three components: (i) Search Space, (ii) Optimization Method, and (iii) Evaluation Strategy. The search space comprises possible Neural Network (NN) architectures for exploration. For example, in the case of CNN-based NAS, a search space consists of a set of operations used in convolution blocks (convolution, fully connected, pooling layers), and numerous architectures are formed based on their combinations. Furthermore, the combination of these layers is represented either by a Direct Acyclic Graph (DAG) encoding or some meta-architecture representation. The optimization method also called the search algorithm, explores the search space to find the optimal architecture. Over the last decade, Automatic Deep Learning (AutoDL) and NAS have gained popularity due to their ability to solve various problems automatically. The evaluation strategy evaluates the performance of the search architecture on the training dataset. One of the simple evaluation strategies is to train the neural network from scratch, but to reduce the evaluation time, multiple approaches have been proposed, such as early stopping, surrogate-based, one-shot, and zero-shot proxy-based methods.

Following the literature, multiple methods have been proposed to search for best performing architecture from search space, such as Reinforcement Learning (RL), Random Search (RS), Bayesian Optimization (BO), gradient-based optimization, and Evolutionary Approaches (EA). Early research on NAS utilizes RL-based algorithms. However, the computational time of this approach is very long [9]. Gradient-based methods are more efficient compared to RL methods. Unfortunately, they often find ill-conditioned architectures and require constructing supernet architecture in advance. This latter needs human expertise to initially design supernet and makes the approach semi-automatic [4]. Multiple evolutionary approaches have been proposed to search for an architecture for a given problem. Due to their ability to self-adopt the search for the optimum solution using nature-inspired computing they are widely adopted for NAS. Genetic Algorithm (GA) is a popular metaheuristic algorithm inspired by the theory of natural selection process. Metaheuristics have been found effective in searching for Deep Learning (DL) architectures in multiple application domains such as image classification [5,8], time series classification [6] and medical Image segmentation [7] etc.

Various studies have been proposed in the literature for medical image classification using NAS. However, these studies have considered only a limited number of datasets and performance measures. In order to overcome these issues, in this study an approach is proposed for designing CNN architecture using GA for medical image classification. The contribution of proposed GAMED-A-CNN approach is as follows:

- A GA approach is proposed for designing attention-based CNN architectures for medical image classification that use a variable length encoding scheme to explore a diverse population of individuals consisting of different number of layers.
- We performed experiments on multiple benchmark datasets, including MedMNISTv2, as well as breast, chest, and brain datasets. These experiments involved evaluating performance measures across different datasets.
- A monoobjective fitness function is introduced to combine multiple objective functions using weighted sum approach.
- Extended the proposed approach for weakly-supervised segmentation by generating heat-maps from Grad-CAM layer to verify the reliability of the proposed approach.

The rest of the article is organized as follows: The related work is discussed in Sect. 2, In Sect. 3, the proposed GA-CNN approach with the explanation of sub-modules in the proposed approach is presented. In Sect. 4, the results and experimental settings are discussed in details. Finally the paper is concluded in Sect. 5.

2 Related Work

Several studies have been proposed for designing CNN-based architectures using GA and other metaheuristics [8,10,11]. These studies use two types of search structures: (i) micro and (ii) macro structure. In macro structure, NN's topological structure is built by finding connections between cells, whereas the cells consist of convolution, pooling, and other layers. The optimization algorithm searches for the operations between the cells or nodes inside the neural network in micro structure.

Usually, a CNN architecture consists of multiple building blocks, also known as layers, i.e., convolution, pooling, attention, and normalization layers. These layers contain parameters that need to be optimized, such as the number of filters, kernel, and padding size in the convolution layer, pooling size in the max-pooling layer, and probability of dropping the neural units in the dropout layer. The convolution layer uses the kernels to perform convolution operations on an image viewed as a matrix. Multiple convolution layers stacked one after another with variable channel and kernel sizes assist CNN in automatic feature extraction and more refined features. The pooling layer down-samples the feature maps by selecting the maximum, minimum, or average value in each patch of the feature map. This patch is also called a grid of size [gxg], where the value of g is a parameter. To overcome the overfitting problems faced by CNN architectures, multiple approaches have been used. Skip connections are found to be effective in overcoming the overfitting problem [15]. ResNet architecture based on residual connections outperforms previous architectures [15]. However, these architectures are human designed and the building of these architectures requires human expertise. To overcome this issue, NAS algorithms try combinations of skip connections between different layers and search for the best performing architecture.

Recently, attention mechanisms have shown state-of-the-art performance on Natural Language Processing (NLP) tasks. Motivated by attention in NLP [12], researchers proposed to use visual attention for computer vision tasks [13]. The attention mechanism is inspired from the human visual system, which naturally finds salient regions in complex scenes. Such a system adopts dynamically weight adjustment process based on the image features. The attention mechanism in computer vision can be treated as a dynamic selection process that adaptively assign weights to features according to the importance of input. Only limited number of studies have used NAS for searching attention-based CNN architectures for computer vision tasks [14].

A large number of studies have been proposed for designing CNN architectures using GA. In [8], the authors proposed a GA-based approach that searches for CNN architectures consisting of skip connections and multiple convolution layers using a variable size encoding scheme. Their proposed approach achieves satisfying results on the CIFAR 10 dataset. Similarly, in [16], the authors proposed an automatic CNN design approach that uses a novel chromosome representation scheme designed to achieve high accuracy within limited resources. To improve the efficiency of the proposed approach, they adopted an ensemble-based majority voting approach. In the first step, multiple CNN models are generated using the GA approach. Then, an ensemble of top-performing individuals is built to achieve better accuracy.

Other studies adopt CNN for the optimization of hyperparamters. Johnson et. al [17] proposed a sequential crossover operation, using an incremental selective schedule that leads to higher diversity in early generations and evaluating individual performance with early stopping to reduce the evaluation and searching time [17]. An evolution-grammar approach is proposed for designing CNN architecture for medical image classification in [18]. They formulated the problem as a grammar representing the encoding of CNN architecture and multi-objectives for fitness evaluation. They evaluated the performance of the proposed approach on three datasets from the MedMNIST benchmark [3][1].

The encoding scheme and fitness evaluation plays an important role while designing CNN architectures using GA [19]. The encoding scheme consists of multiple steps. In each step, a small architecture is encoded and different small architectures are stacked to build more complex architecture. In [8], the authors show that using variable length encoding is better compared to the fixed length

---

[1] MedMNIST v2: A Large-Scale Lightweight Benchmark for 2D and 3D Biomedical Image Classification. available at https://medmnist.com/.

one because the optimal depth of CNN architecture is unknown for a given problem. Inspired by this work, a variable length encoding approach is adopted in our study.

3 Proposed GAMED-A-CNN Approach

The proposed GAMED-A-CNN algorithm consists mainly of three steps (i) population initialization (ii) recombination and crossover operations and (iii) fitness evaluation. The graphical representation of proposed methodology is shown in Fig. 1 and algorithm is shown in Algorithm 1, in which all the steps of proposed methodology are summarized.

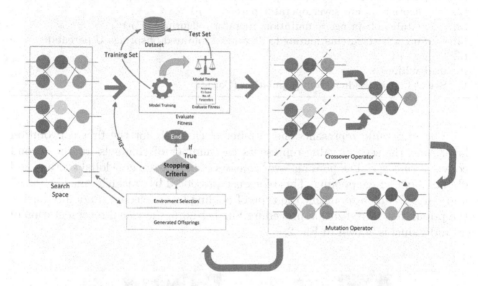

Fig. 1. Graphical Representation of Proposed Methodology

3.1 Representation of Individual

The CNN architecture generated by the proposed algorithm combines multiple blocks using skip connection to generate different architectures. These blocks consist of Convolution Attention Skip Layers (CASK) block, Pooling Layers (PL) block and a Fully Connected (FC) layer. The CASK block consists of two convolution layers, with a batch normalization layer and a Channel-Wise Attention (CWA)-2D layer after the second convolution layer. The CWA-2D layer is used [25] which performs attention on channels of the previous layers. In CWA, to reduce the number of operations the number of channels are reduced by applying 1D convolution on the input. A 128-32 string represents the CASK block.

Algorithm 1. Proposed GA-CNN for Medical Images Classification

---

1: **Input**: Number of Epochs (N_E), Crossover Probability (C_P), Mutation Probability (M_P), Dataset (D), Population size (P_S), Number of Generations (NoG)

2: **Output**: Discovered best architecture

3: P_0: Initialize Population randomly with given Population size (P_S)

4: g:= 1

5: **while** g $\leq$ NoG **do**

6:      model = decode_model(g,Pop)

7:      model = Train model on dataset D

8:      Accuracy,F1-Score,Parameters = Evaluate model on dataset D

9:      Fitness = α* Accuracy + β * F1 Score +(1 - γ) * log (Parameters)

10:     Store individual in population with obtained fitness

11:     parents = Select individual parents from population Pop

12:     offsprings = crossover_operator(parents, C_P)

13:     mutated_offsprings = mutation_operator (offsprings, M_P)

14:     Pop = best_parents_mutated_offsprings (mutated_offsprings U parents)

15:     g=g+1

16: **end while**

17: Select best individual from population (Pop)

The first value represents the number of channels for the first convolution layer, and the second value represents the number of channels for the second convolution layer. The pooling layer consists of max, average, global average and global max average pooling. This block is represented by "mean", "max", "gavg" and "gmax" which represent the type of pooling layer selected randomly during the population initialization procedure.Furthermore, the visual representation of an individual is shown in Fig. 2.

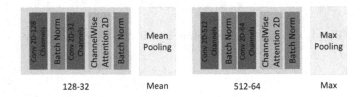

Fig. 2. Graphical representation of individual with 128-32-mean-512-64-max phenotype.

3.2   Population Initialization

Population initialization is the first phase of GAMED-A-CNN algorithm in which the individuals are generated randomly. During initialization, the length of an individual is represented by L, which is initialized randomly. The individual is stored in a linked list data structure containing L nodes, whereas each node represents some layer or block. Linked-list is adopted for this problem because it is dynamic and handles variable length input.

3.3   Fitness Function

The fitness function used to evaluate individuals uses a weighting method to combine multiple performance measures (accuracy, f1-score, and the number of parameters). The key idea is to combine the maximum objective (accuracy and f1-score) and minimum objective (number of parameters) using a weighted sum approach. The goal is to identify the individual with the maximum accuracy and precision score with fewer parameters. The fitness function is formulated as:

$$\text{Fitness} = \alpha * \text{Accuracy} + \beta * \text{F1 Score} + (1 - \gamma) * \log(\text{Parameters})$$

The values of these parameters adopted in this study are: $\alpha = 0.4$, $\beta = 0.4$ and $\gamma = 0.8$. The key idea is to have the sum of weights equal to 1. These values are obtained after experimenting with different values of α, β and γ, as these values directly affects the quality of the individuals obtained.

3.4   Crossover Operation

Three-point crossover is adopted in this study to generate individual offsprings from parents. Compared to a two-point crossover approach, a three-point crossover approach results in diversity among individuals, and offspring have diverse representation from both individuals. If the generated random number is greater than the crossover probability, which is 0.9 by default as suggested in [26], the crossover operation is not performed. This probability value allows maximum individuals to be generated and added to the population. For crossover operation, two individuals are selected from the population list. The selection criteria for these individuals are based on their fitness values. A three-point crossover is then performed on these two individuals to produce offspring as illustrated in Fig. 3.

Fig. 3. Graphical representation of crossover operator.

3.5   Mutation Operation

The mutation operation is applied to the selected individual by morphing the phenotype of an individual. The mutation operation does not occur if the generated random number exceeds the mutation probability, which is 0.2 as suggested in [26] by default. At first, a layer is randomly selected from the individual representation, consisting of multiple layers. Then one operation is randomly selected and replaced from three operations (i) adding skip or pooling layer (ii) changing layer with another layer (iii) removing the layer.

4   Results and Experimental Settings

4.1   Experimental Settings

The proposed GAMED-A-CNN algorithm is implemented in Python and Keras, a DL framework. The total number of generation sizes is set to 20 with a population size of 20. For comparison with multiple DL architectures like ResNet, Inception, Xception, etc., we extended the implementations provided by the MedMNIST authors in the PyTorch framework. For experimentation, the same parameter setting is adopted. A learning rate of 0.001 with Adam optimizer is used with batch size of 64 for 100 epochs. All the experiments have been conducted on NVIDIA GeForce GTX 1080 Ti GPU.

Nine datasets from the MedMNIST benchmark belonging to multiple organs and modalities is used. These datasets include colon pathology, breast mammograms, dermatology images, lung nodules x-rays, and multiple organ datasets. Furthermore, three other high-resolution datasets are also used for this study. The dataset of breast mammogram masses consists of a combination of three different breast mammogram datasets, namely: INbreast, MIAS and DDSM, which are three famous mammograms [20]. For brain Magnetic Resonance Imaging (MRI) scans, The images from DICOM scans are split into tumor and non-tumor from Brain Tumor Segmentation (BRaTS) 2019 dataset [24].

4.2   Results and Discussion

A number of experiments have been conducted to evaluate the proposed approach. Several performance measures are used to compare the proposed approach with existing DL architectures, including accuracy, F1-score, and Area Under the Curve. (AUC). An ablation study is also performed on the proposed approach, comparing the effects of using different population sizes on multiple datasets. Experiments on different population sizes of 10, 15, and 20 are performed on multiple datasets , as shown in Table 1. It is observed that the increase in population size directly affects the performance of the proposed approach. As the large population size directly affects the exploration of good individuals among large population size.

Table 1. Results on different variations of GA-CNN approach.

Performance Measure	Type of Model	Dataset Name									
		Adresnal	Blood	Breast	Derma	Nodule MNIST3D	Organ A MNIST	Organ C Mnist	Organ MNIST 3D	Path MNIST	Synapse MNIST 3D
ACCURACY	GA CNN- 10	0.82	0.94	0.87	0.72	0.91	0.92	0.87	0.83	0.83	0.72
	GA-CNN-15	0.82	0.95	0.88	0.73	0.93	0.92	0.9	0.83	0.85	0.76
	GA-CNN-20	**0.83**	0.95	**0.89**	**0.77**	0.93	**0.95**	**0.92**	0.76	**0.93**	**0.77**
F1 SCORE	GA CNN- 10	0.57	0.92	0.9	0.71	0.7	0.92	0.85	0.84	0.83	0.81
	GA-CNN-15	0.57	**0.95**	0.92	0.72	**0.71**	0.92	0.9	**0.87**	0.85	0.85
	GA-CNN-20	**0.59**	0.84	0.92	**0.76**	0.64	**0.95**	**0.92**	0.76	**0.93**	**0.86**
AUC	GA CNN- 10	0.75	0.97	0.88	0.94	0.91	0.99	0.89	0.98	0.95	0.69
	GA-CNN-15	0.84	**0.99**	0.89	0.95	0.93	0.99	0.99	**0.99**	0.97	0.66
	GA-CNN-20	0.84	0.98	**0.9**	0.95	0.93	0.99	0.99	0.94	**0.98**	**0.72**
Number of Parameters	GA CNN- 10	1451457	3674167	3278432	5128661	896531	735915	5687843	7923861	5213183	**434521**
	GA-CNN-15	**158977**	2647176	**3169025**	4414343	**696609**	**627915**	**5416843**	4816683	**3076265**	442401
	GA-CNN-20	5207489	479816	4443649	1887719	4889889	1060427	4098699	660747	4440521	991873

In Table 3 the proposed algorithm is compared with DL architectures (VGG16, VGG19, ResNet 50, ResNet101, Xception and InceptionV3). Compared to most algorithms, the proposed method performs better on all performance measures and has fewer parameters than existing DL architectures. In addition, it is observed that after embedding the CWA-2D in the individual block, some of the networks generated by GA achieved higher f1 scores and AUC scores than GA-CNN and some achieve lower scores, which means the attention layer somehow assists the network in having more positive predictions but sometimes it fails to find a suitable architecture with attention layer due to fixed population size. The nature of attention is to dynamically adjust the weights according to the importance of the input image.

However, compared to the number of parameters, the networks generated by the proposed approach contain up to 50% fewer parameters than the ResNet architecture.The p-values of accuracy, f1-score, and AUC are given below with the null hypothesis that these methods have no difference. The friedman test is used to first reject the null hypothesis at the significance level of 0.05 for evaluation of results statistically. The value of $p < 0.05$ means a significant difference between the methods.The results of the proposed approach is also compared with AutoML techniques mentioned in the MedMNIST article [3] named AutoSKlearn and AutoKeras. In AutoKeras, the authors proposed a NAS approach based on network morphism by searching the architecture and hyperparameters with the bayesian optimization approach. Their approach uses a network kernel, and tree-structured acquisition function in bayesian optimization for efficient exploration of search space [22].

Table 2. Comparison of proposed GA-CNN approach with AutoKeras and AutoSKlearn AutoML approaches.

Type of Model	Dataset																		
	Blood		Breast		Derma		Organ A MNIST		Organ C Mnist		Path MNIST		TissueMNIST		Pnumenia MNIST		OrganSMNIST		OCT MNIST	
	ACC	AUC	ACC	AUC	ACC	AUC	ACC	AUC	ACC	AUC	ACC	AUC	ACC	AUC	ACC	AUC	ACC	AUC	ACC	AUC
GA-CNN-20	0.95	0.98	0.89	0.9	**0.77**	0.95	**0.95**	0.99	**0.92**	0.99	**0.93**	**0.98**	0.55	0.88	0.86	0.92	0.93	0.87	0.78	0.89
GA-CNN-Attention	0.96	**0.99**	**0.91**	**0.92**	0.72	0.95	0.87	0.97	0.74	0.97	0.74	0.95	0.67	0.92	0.9	0.95	**0.94**	**0.98**	0.77	0.94
AutoKeras [3]	0.87	0.98	0.8	0.87	0.71	0.91	0.76	0.99	0.82	0.99	0.71	0.95	0.82	**0.94**	0.85	0.94	0.67	0.97	0.78	**0.95**
Auto-Sklearn [3]	0.96	0.98	0.83	0.83	0.74	0.9	0.9	0.96	0.87	0.97	0.83	0.93	**0.94**	0.82	0.87	0.94	0.81	0.94	**0.87**	0.87

AutoSKlearn formulates the problem as Combined Algorithm Selection and Hyper-parameter Optimization (CASH). They used two components for hyper-parameter optimization: (i) meta-learning to initialize the Bayesian optimizer (ii) automatic ensemble construction from configurations evaluated during optimization [21]. Table 2 compares the proposed approach with AutoSKlearn and AutoKeras in terms of AUC and accuracy scores. The AUC and accuracy scores of the proposed approach outperform in most of the datasets. The AutoSKlearn and AutoKeras took more time to search for an optimal architecture than the proposed approach.

Furthermore, the line graph as shown in Fig. 4 visualize the comparison of CNN parameters generated by the proposed approach with simple accuracy and weighted-sum based fitness function . The parameters generated using the simple accuracy fitness functions are larger than the modified fitness function, which means the modified fitness function assists the proposed approach in finding architectures with fewer parameters.

Table 3. Comparison of proposed approach with existing deep learning approaches.

Performance Measure	Dataset									
	Type of Model	Blood	Breast	Derma	Organ A MNIST	Organ C Mnist	Path MNIST	TissueMNIST	Pnumenia MNIST	OrganSMNIST
ACCURACY	vgg16	0.87	0.82	0.72	0.88	0.86	0.91	0.63	0.84	0.74
	vgg19	0.95	0.82	0.67	0.94	0.87	0.63	0.62	0.86	0.75
	resnet18	0.95	0.87	0.74	0.94	0.91	0.91	0.66	0.83	0.79
	resnet50	0.95	0.79	0.73	0.83	0.9	0.9	0.67	0.87	0.78
	resnet101	0.95	0.73	0.73	0.94	0.9	0.92	0.66	0.87	0.78
	Inceptionv3	0.95	0.75	0.74	0.92	0.89	0.92	0.66	0.87	0.79
	xception	0.94	0.78	0.74	0.94	0.91	0.9	0.66	0.85	0.79
	GA-CNN-20	0.95	0.89	0.77	0.95	0.92	0.93	0.55	0.86	0.94
	GA-CNN-Attention	0.96	0.91	0.72	0.87	0.74	0.74	0.67	0.9	0.94
	p-Value	1.77E-06	6.42E-09	1.15E-10	2.87E-08	4.28E-08	0.00048084	3.71E-09	3.09E-05	3.90E-05
F1 SCORE	vgg16	0.87	0.81	0.71	0.88	0.86	0.9	0.63	0.83	0.74
	vgg19	0.95	0.82	0.54	0.94	0.87	0.63	0.62	0.96	0.75
	resnet18	0.95	0.87	0.73	0.94	0.91	0.91	0.62	0.82	0.79
	resnet50	0.95	0.78	0.71	0.84	0.9	0.9	0.61	0.86	0.78
	resnet101	0.95	0.84	0.71	0.94	0.9	0.91	0.66	0.86	0.78
	xception	0.92	0.86	0.73	0.94	0.91	0.9	0.65	0.89	0.79
	Inceptionv3	0.95	0.85	0.73	0.92	0.89	0.92	0.66	0.87	0.79
	GA-CNN-20	0.84	0.92	0.76	0.95	0.92	0.93	0.54	0.9	0.92
	GA-CNN-Attention	0.96	0.89	0.72	0.87	0.74	0.74	0.63	0.92	0.93
	p-value	2.79E-07	4.03E-11	8.84E-08	2.50E-08	4.34E-08	0.00058632	2.78E-08	6.87E-14	1.19E-05
AUC	vgg16	0.98	0.85	0.87	0.99	0.98	0.99	0.89	0.95	0.96
	vgg19	0.98	0.85	0.92	0.99	0.98	0.93	0.89	0.93	0.96
	resnet18	0.99	0.89	0.9	0.99	0.99	0.98	0.89	0.93	0.97
	resnet50	0.99	0.82	0.88	0.98	0.99	0.97	0.91	0.95	0.97
	resnet101	0.99	0.57	0.88	0.99	0.98	0.99	0.91	0.82	0.97
	Inceptionv3	0.99	0.75	0.89	0.96	0.98	0.99	0.91	0.94	0.97
	xception	0.98	0.75	0.97	0.99	0.98	0.97	0.91	0.92	0.97
	GA-CNN-20	0.98	0.9	0.95	0.99	0.99	0.98	0.88	0.92	0.87
	GA-CNN-Attention	0.99	0.92	0.95	0.99	0.99	0.92	0.92	0.95	0.98
	p-value	1.33E-10	3.34E-07	9.00E-15	1.69E-14	1.37E-13	9.88E-07	3.40E-08	3.49E-11	8.17E-07
Number of Parameters	vgg16	82200000	82200000	82200000	82200000	82200000	82200000	82200000	82200000	82200000
	vgg19	138,000,000	138,000,000	138,000,000	138,000,000	138,000,000	138,000,000	138,000,000	138,000,000	138,000,000
	resnet18	11,511,784	11,511,784	11,511,784	11,511,784	11,511,784	11,511,784	11,511,784	11,511,784	11,511,784
	resnet50	25,600,000	25,600,000	25,600,000	25,600,000	25,600,000	25,600,000	25,600,000	25,600,000	25,600,000
	resnet101	44,500,000	44,500,000	44,500,000	44,500,000	44,500,000	44,500,000	44,500,000	44,500,000	44,500,000
	Inceptionv3	23,885,392	23,885,392	23,885,392	23,885,392	23,885,392	23,885,392	23,885,392	23,885,392	23,885,392
	xception	22,800,000	22,800,000	22,800,000	22,800,000	22,800,000	22,800,000	22,800,000	22,800,000	22,800,000
	GA-CNN-20	479816	4443649	1887719	1060427	4098699	4440521	1623432	126145	8974651
	GA CNN Attention	1104465	152452	1224714	1894225	1797809	184626	1890766	10180	9572631

In Table 4, the results on three different datasets apart from MedMNIST on proposed GA-CNN is given. It is noted that the proposed approach can find

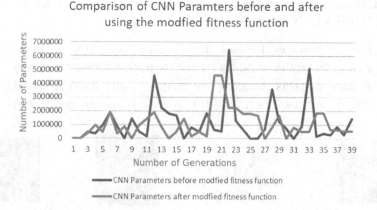

Fig. 4. Comparison of Number of parameters of generated CNN by proposed GA-CNN algorithm using modified fitness function on Blood dataset.

high-performing architecture with a small number of parameters. Class Activation Map (CAM) is a technique to generate heat maps to highlight class-specific regions of interest. CAMs are designed to provide insights into how CNN makes different predictions by highlighting the regions that contribute most to the predicted class. This helps visualize the regions within the image that contribute most while making the decision. Grad-CAM (Gradient-weighted Class Activation Mapping) is a famous visualization technique that uses the gradient of the target concept flowing into the last convolution layer to generate a localization map highlighting the important region for a visual explanation of the concept [23]. The visual heatmaps are generated using the Grad-CAM approach from networks generated by the proposed approach on brain, breast and chest datasets to verify the effectiveness of the proposed approach. In Fig. 5, some of the heatmaps generated from chest, brain and breast datasets are shown. The red region shows that the probability of predictions is very high and highlights the important region. As in our problem, the network task is to identify the tumorous/injuries region in the image. The Figs. 5a and 5d visualize the heatmaps generated from brain MRI scans, highlighting the tumorous region which means the architecture successfully locates the class-specific region. Similarly in Fig. 5b, the breast tumor region is highlighted by red pixel values and Fig. 5c highlights the infected region. These predictions can also be formulated for segmentation problem in which the infected region is located, also called weakly-supervised segmentation.

Table 4. Results of breast density, brain tumor and chest pneumonia datasets classified using GAMED-A-CNN approach.

Dataset	Accuracy	Precision	Recall	F1-Score	AUC	Params
Brain	0.837104	0.781553	0.947059	0.856383	0.909725	12289
Breast	0.733898	0.840035	0.643591	0.728807	0.830682	12289
Chest	0.773305	0.775597	0.995074	0.871733	0.653106	3073

(a) Brain Tumor   (b) Breast Cancer   (c) Pneumonia   (d) Brain Tumor

Fig. 5. Visual Heat-maps generated from multiple chest, brain and breast dataset outlining possible application of proposed approach for semi-supervised segmentation using NAS.

5 Conclusion

This study proposes a genetic algorithm-based approach for searching CNN architecture for medical image classification problems. A variable-length encoding scheme was introduced to represent the architecture and weighted sum fitness function. Numerical and visual experiments on MedMNIST and three other datasets have shown the effectiveness of the proposed approach in terms of different performance measures. In the near future, more applications of NAS for medical image analysis will be explored. Furthermore, our aim is to investigate the influence of other metaheuristics when searching for optimal CNN architectures.

Acknowledgment. This work was funded by ArtIC project "Artificial Intelligence for Care" (grant ANR-20-THIA-0006-01) and co-funded by IRIMAS Institute/Université de Haute Alsace. The authors would like to thank the Mesocentre of Strasbourg for providing access to the GPU cluster as well as the providers of all datasets used in this paper.

References

1. Voulodimos, A., Doulamis, N., Doulamis, A., Protopapadakis, E.: Deep learning for computer vision: a brief review. Comput. Intell. Neurosci. **1**, 2018 (2018)
2. Litjens, G., et al.: A survey on deep learning in medical image analysis. Med. Image Anal. **1**(42), 60–88 (2017)
3. Yang, J., et al.: MedMNIST v2: A large-scale lightweight benchmark for 2d and 3d biomedical image classification. arXiv preprint arXiv:2110.14795. 2021 Oct 27

4. Liu, Y., Sun, Y., Xue, B., Zhang, M., Yen, G.G., Tan, K.C.: A survey on evolutionary neural architecture search. IEEE Trans. Neural Netw. Learn. Syst. **34**, 550–570 (2021)
5. Yotchon, P., Jewajinda, Y.: Hybrid multi-population evolution based on genetic algorithm and regularized evolution for neural architecture search. In: 2020 17th International Joint Conference on Computer Science and Software Engineering (JCSSE) 2020 Nov 4, pp. 183–187. IEEE
6. Rakhshani, H., et al.: Neural architecture search for time series classification. In: 2020 International Joint Conference on Neural Networks (IJCNN), 2020 July 19, pp. 1–8. IEEE (2020)
7. Weng, Y., Zhou, T., Li, Y., Qiu, X.: NAS-UNET: neural architecture search for medical image segmentation. IEEE Access **4**(7), 44247–57 (2019)
8. Sun, Y., Xue, B., Zhang, M., Yen, G.G., Lv, J.: Automatically designing CNN architectures using the genetic algorithm for image classification. IEEE Trans. Cybern. **50**(9), 3840–54 (2020)
9. Zoph B, Le QV. Neural architecture search with reinforcement learning. arXiv preprint arXiv:1611.01578. 5 November 2016
10. Wang, B., Sun, Y., Xue, B., Zhang, M.: A hybrid GA-PSO method for evolving architecture and short connections of deep convolutional neural networks. In: Nayak, A.C., Sharma, A. (eds.) PRICAI 2019. LNCS (LNAI), vol. 11672, pp. 650–663. Springer, Cham (2019). https://doi.org/10.1007/978-3-030-29894-4_52
11. Li, Y., Xiao, J., Chen, Y., Jiao, L.: Evolving deep convolutional neural networks by quantum behaved particle swarm optimization with binary encoding for image classification. Neurocomputing **14**(362), 156–165 (2019)
12. Vaswani, A., et al.: Attention is all you need. In: Advances in Neural Information Processing Systems, vol. 30 (2017)
13. Guo, M.H., et al.: Attention mechanisms in computer vision: a survey. arXiv preprint arXiv:2111.07624. 15 November 2021
14. Fan, Z., Hu, G., Sun, X., Wang, G., Dong, J., Su, C.: Self-attention neural architecture search for semantic image segmentation. Knowl. Based Syst. **5**(239), 107968 (2022)
15. He K, Zhang X, Ren S, Sun J. Deep residual learning for image recognition. In: Proceedings of the IEEE Conference on Computer Vision and Pattern Recognition 2016, pp. 770–778 (2017)
16. Ahmed, A.A.F., Darwish, S.M.S., El-Sherbiny, M.M.: A novel automatic CNN architecture design approach based on genetic algorithm. In: Hassanien, A.E., Shaalan, K., Tolba, M.F. (eds.) AISI 2019. AISC, vol. 1058, pp. 473–482. Springer, Cham (2020). https://doi.org/10.1007/978-3-030-31129-2_43
17. Johnson, F., Valderrama, A., Valle, C., Crawford, B., Soto, R., Nanculef, R.: Automating configuration of convolutional neural network hyperparameters using genetic algorithm. IEEE Access. **25**(8), 156139–156152 (2020)
18. da Silva, C.A., Miranda, P.B., Cordeiro, F.R.: A new grammar for creating convolutional neural networks applied to medical image classification. In: 2021 34th SIBGRAPI Conference on Graphics, Patterns and Images (SIBGRAPI), 2021 October 18, pp. 97–104. IEEE (2021)
19. White, C., Neiswanger, W., Nolen, S., Savani, Y.: A study on encodings for neural architecture search. Adv. Neural. Inf. Process. Syst. **33**, 20309–20319 (2020)
20. Huang, M.L., Lin, T.Y.: Dataset of breast mammography images with masses. Data Brief **1**(31), 105928 (2020)

21. Feurer, M., Klein, A., Eggensperger, K., Springenberg, J., Blum, M., Hutter, F.: Efficient and robust automated machine learning. In: Advances in Neural Information Processing Systems, vol. 28 (2015)
22. Jin, H., Song, Q., Hu, X.: Auto-keras: an efficient neural architecture search system. In: Proceedings of the 25th ACM SIGKDD International Conference on Knowledge Discovery & Data Mining, 2019 July 25, pp. 1946–1956 (2019)
23. Selvaraju, R.R., Cogswell, M., Das, A., Vedantam, R., Parikh, D., Batra, D.: Grad-CAM: visual explanations from deep networks via gradient-based localization. In: Proceedings of the IEEE International Conference on Computer Vision 2017, pp. 618–626 (2017)
24. Menze, B.H., et al.: The multimodal brain tumor image segmentation benchmark (BRATS). IEEE Trans. Med. Imaging 34(10), 1993–2024 (2014)
25. Wang, Q., Wu, B., Zhu, P., Li, P., Zuo, W., Hu, Q.: ECA-Net: efficient channel attention for deep convolutional neural networks. In: IEEE International Conference on Computer Vision and Pattern Recognition, pp. 11531–11539 (2020)
26. Back, T.: Evolutionary algorithms in theory and practice: evolution strategies, evolutionary programming, genetic algorithms. Oxford University Press, 11 January 1996

A Multi-objective 3D Offline UAV Path Planning Problem with Variable Flying Altitude

Mahmoud Golabi[1]([envelope]) [iD], Soheila Ghambari[1] [iD], Shilan Amir Ashayeri[2],
Laetitia Jourdan[3] [iD], and Lhassane Idoumghar[1] [iD]

[1] Université de Haute-Alsace, IRIMAS UR 7499, 68100 Mulhouse, France
{mahmoud.golabi,soheila.ghambari,lhassane.idoumghar}@uha.fr
[2] Mulhouse, France
[3] University of Lille, CRIStAL, UMR 9189, CNRS, Centrale Lille, France
laetitia.jourdan@univ-lille.fr

Abstract. This study focuses on a 3D multi-objective collision-free offline UAV path planning problem by considering the variability in flying altitude over an urban environment that is replete with static obstacles. The environment is decomposed into several equal-sized ground cells with an infinite flying altitude. The UAV can adjust the altitude to fly above obstacles or bypass them by choosing another way. Various cells may have different flying risks at different altitude levels. This study aims to find the most efficient and safe trajectory toward the destination while maximizing the number of visited obstacle-free cells. This aim could be followed in real-world surveillance and disaster tracking operations, where the users may want to collect data even en route to the destination. This problem is formulated as a multi-objective mixed-integer non-linear mathematical model in which minimizing the flying distance, the required energy, the maximum path risk, and the number of not-visited obstacle-free cells are the objective functions. An enhanced Non-dominated Sorting Genetic Algorithm-II (NSGA-II) is developed to solve the problem. Computational results indicate the superiority of the developed algorithm in solving several real-world data sets.

Keywords: Path Planning · UAV · Multi-objective Optimization · NSGA-II

1 Introduction

Unmanned Aerial Vehicles (UAVs), also known as drones, are aerial vehicles without carrying a human operator. They could be controlled remotely or fly automatically. The first purpose of designing UAVs was to deploy them in military operations. Something that has led to the era of drone wars regarding the significant role of armed UAVs in Libya, Nagorno-Karabakh, Syria, and Ukraine wars [6]. Amazon's Prime Air project in 2013 was a milestone in using UAVs for

---

S. S. Ashayeri—Independent Scholar.

P. Legrand et al. (Eds.): EA 2022, LNCS 14091, pp. 187–200, 2023.
https://doi.org/10.1007/978-3-031-42616-2_14

civil missions. Labor cost reduction, higher moving speed, wider aerial vision, and avoiding traffic jams are the main advantages of aerial vehicles. UAVs flexibility and cost efficiency have made them one of the most popular aerial vehicles for a wide variety of civil operations such as last-mile delivery [25], relief distribution [13], mapping [29], search and rescue [3], surveillance [19], etc.

The mentioned extended scope of applications highlighted the need for optimization studies in UAV-based missions. The literature is now replete with optimization studies focusing on determining the best location of launching and/or refueling stations [26], the optimal visit sequence and route of drones [21], the shortest path to the target point [11], etc. This study aims to focus on the UAV path planning problem by introducing a new multi-objective version that tries to maximize the number of en route visited cells while minimizing the path length, the required energy, and minimizing the aggregation of the maximum path risk.

As one of the main problems in the navigation of autonomous UAVs, path planning aims to safely direct each UAV toward the desired destination such that the main goals and objectives are fulfilled [14]. Generally, the main objective is to find the shortest collision-free path, however, other objectives such as optimizing the required energy, path smoothness, path risk, and the number of visited cells are also considered. The main objective could be considered as a criterion to categorize path planning problems based on the applications. The main type is the UAV shortest path problem which aims to find the shortest traveling distance to reach the destination. Considering the available resources, informative path planning maximizes the collected data about an unknown environment [22]. Cooperative path planning refers to coordinated missions having a set of UAVs flying simultaneously [2]. Determining a path that passes through all points of the area of interest refers to the full coverage path planning problem [15]. Maximum coverage path planning [5] could be a solution for cases in which limited resources are not enough to cover the entire area. It is noteworthy to mention that there are some multi-objective studies considering a combination of these objectives, simultaneously. The existing literature surveys [28,33,34] scrutinize different types of UAV path planning problems and corresponding solution algorithms. Path planning could also be categorized into offline and online problems [31]. Offline path planning is applied in a completely known environment, where the best path could be obtained prior to the actual flight. Online path planning is for uncertain environments with unknown moving obstacles and adversaries.

Modeling the environment is one of the key issues in UAV path planning problems. The 3D flying space, known as world space, has been represented using different methods in the literature [8]. Voronoi diagrams were among the first methods applied to represent free space using a network of one-dimensional lines. The cell decomposition approach represents the free space with a set of convex polygons, called cells. The adjacency relationship among the cells is represented by the connectivity graph. The potential field is another method that represents the free space by assigning repulsive and attractive forces to obstacles and the destination, respectively, such that the UAV can move toward the goal in a collision-free manner. One of the predominantly used methods to represent the environment is the occupancy grid map that decomposes the world space into a set of equal-sized cells with unique indices. The cells could be related to the free space or obstacles.

Based on the type of the studied problem and the characteristics of the applied environment modeling technique, there may be different methods to solve the UAV path planning problem [30]. Like the majority of optimization problems, the path planning problem can be formulated as a mathematical model and solved using the existing mathematical approaches. The complexity of the studied problem and the size of the solution space are the main barriers to solving the developed mathematical models. Graph search methods are among the classical techniques to solve the UAV path planning problem. The Dijkstra and A* algorithms are the most well-known graph search methods applied to find the shortest paths. Among classical methods, A* is well-known, as the most efficient approach in larger world spaces [4]. Sampling-based methods such as probabilistic roadmap (PRM) and rapidly exploring random tree (RRT) are effective planners to find feasible solutions in a short time, without focusing on optimality conditions. The potential field is another method that not only represents the space but also determines the collision-free path to reach the destination. Although this method has a quick response speed, it can be easily trapped in local minima, especially in larger environments. These solution approaches are generally suitable for single-objective path planning problems. They are mainly focused on finding the shortest collision-free path between origin and destination. Introducing new objective functions or studying the problem in a multi-objective framework requires implementing other solution algorithms. Metaheuristics are the solution algorithms that can be applied to solve these types of path planning problems in both single or multiple UAV scenarios. By a glimpse into the literature one can find several applications of metaheuristics such as Genetic Algorithms [27], Ant Colony Optimization [16], and Differential Evolution [32] in solving UAV path planning problems.

The majority of UAV path planning problems are inspired by robotic motion planning studies in which moving through an area occupied by an obstacle is forbidden. However, the main difference between a ground-moving robot and a UAV is the possibility of adjusting the altitude and flying above existing obstacles. One of the first studies considering the possibility of flying above obstacles or bypassing them is done by Golabi et al. [12]. Using occupancy grid maps to represent the environment, they studied a multi-objective UAV path planning problem to minimize the path length, the consumed energy, and the path risk, simultaneously. Introducing several constraints and decision variables for linearizing, they formulated the problem as a mixed-integer linear programming model. They solved the problem using several state-of-the-art evolutionary multi-objective optimization algorithms.

The current paper tries to study a maximum-covering version of [12] by adding a new objective function to minimize the number of non-visited obstacle-free cells. Considering a continuous flying altitude level, the problem is formulated using a modified mathematical programming model. Finally, the problem is solved on several real-world 3D datasets using a modified solution algorithm. The rest of this paper is organized as follows. The problem definition is given in Sect. 2. The developed mathematical formulation is presented in Sect. 2.2. Section 3 describes the applied solution algorithms. The obtained results and discussion are presented in Sect. 4. Finally, the paper is concluded in Sect. 5.

2 Problem Definition

2.1 Problem Description

This work studies a 3D UAV path planning problem to find the most efficient and safe trajectory toward the destination while maximizing the number of visited obstacle-free cells. This aim could be followed in real-world surveillance and disaster tracking operations where the users may want to collect data even en route to the destination. The main objectives of this study could be listed as minimizing the path length, the consumed energy, the path maximum risk, and the number of non-visited obstacle-free cells. Adverse to the majority of existing UAV path planning problems that consider a fixed flying altitude level, this work reflects a more realistic image of the problem by taking the possibility of altitude change into account. Traditionally, obstacles are considered as forbidden flying zones in UAV path planning problems. This assumption has been inherited from robotic motion planning studies in which a robot can't move through an area occupied by obstacles. Therefore, bypassing an obstacle and selecting a free way-point is the only option to reach the destination without any collision with obstacles. Adjusting the altitude to fly above the obstacles is another option that could be chosen in the case of having the possibility of changing the flying altitude while moving toward the destination.

Using the occupancy grid map technique, the environment is represented as several equal-sized ground cells with an infinite flying altitude. These cells form a collection of rows and columns on the ground space. The cells occupied with obstacles with the heights of corresponding obstacles are determined in advance. The UAV starts the mission from the origin by following a path to reach the destination such that the considered objectives are optimized. The path could be modeled as a combination of visited ground cells and the concomitant flying altitudes above while passing through them. For each ground cell, the next move is defined using a set of succeeding cells by considering flying in 5 different directions such that the UAV would be either one column closer to the destination, or at least at the same flying column (see Fig. 1). If the selected succeeding cell contains an obstacle, the UAV needs to adjust the altitude level and fly above that obstacle. Therefore, the flying distance between any two consecutive cells would be a function of their ground distance and the amount of altitude change. It should be noticed that the altitude level while flying above different cells may be restricted by air space authorization due to different reasons such as the proximity to military bases, airports, or traditional flying corridors. Based on the previous data and proximity to dangerous obstacles and windy areas,

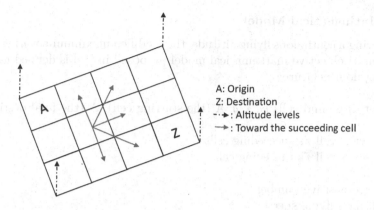

Fig. 1. The flying environment.

there would be various risk factors associated with different altitude levels of each ground cell.

As it was mentioned earlier, the flying distance between any two consecutive cells depends on their ground distance and the change in altitude level. Having these two values, the flying distance is calculated using the Pythagoras theory. The energy consumption will also be a function of both altitude level and flying distance. The consumed energy while moving from point i to point j at different altitude levels is calculated using the formula proposed by Dorling et $al.$ [10]. Since it is assumed that the flying altitude is not fixed, the term ($Wg\Delta_{ij}^+$) is added to the mentioned formulation for considering the extra energy needed to increase the flying altitude:

$$\theta = W^{3/2}\sqrt{\frac{g^3}{2\rho\zeta n}\frac{d_{ij}}{\nu}} + Wg\Delta_{ij}^+ \qquad (1)$$

Here, W is the drone and battery weight (kg), g is the gravity (N), ρ is the fluid density of air (kg/m^3), d_{ij} is the distance between point i and j (m), Δ_{ij}^+ is the increased altitude (m), ν is the flying speed (m/s), ζ is the area of spinning blade disc in m^2, and n is the number of rotors. Considering the normal flying altitude of UAVs, the gravity could be considered as a fixed value as the gravitational force would have decreased just by 1.2% if an object flies at an altitude of 40 km. The fluid density of air is calculated using Eq. 2 [20], where H is the average altitude while flying from point i to point j.

$$\rho = (1 - 2.2558.10^{-5}H)^{4.2577} \qquad (2)$$

For each visiting cell, the maximum risk factor of traversed altitude levels while passing to the successive cell is calculated. The accumulated maximum risk factor is considered as the maximum path risk. On the premise that obstacles are barriers to collecting data about occupied cells, the number of visited obstacle-free cells could be a criterion to optimize the data collected en route to the destination. In other words, it is not possible to observe the area concealed by the obstacles.

2.2   Mathematical Model

Considering a continuous flying altitude, the modified maximum-covering version of the multi-objective mathematical model proposed in [12] is defined using the following nomenclature:

Sets:

I: the set of ground cells (i, j, g, A (the starting cell), N (the final destination) $\in I$)

δ_i^+: the set of cell i's succeeding cells

δ_i^-: the set of cell i's preceding cells

Scalars:

M : a large positive number

v: the drone's flying speed

θ: a scalar used in calculation of consumed energy ($\theta = W^{3/2}\sqrt{g^3/2\zeta n}$)

Parameters:

α_i: 1 if there exist an obstacle at cell i; 0 otherwise

a_j: the height of obstacle located in cell j, if exists

U_j: the maximum allowed flying altitude over cell j

r_{ij}^k: the risk factor concomitant to flying on cell i at altitude k

h^A: the drone's starting altitude

d_{ij}: the direct distance between the neighbour cells i and j

Decision Variables:

X_{ij}: 1 if drone enters cell j from cell i; 0 otherwise

h_{ij}: the adjusted altitude while entering cell j from cell i ($h_{ij} \geq 0$)

Δ_{ij}: absolute value of altitude change while flying from cell i to cell j ($\Delta_{ij} \geq 0$)

Δ_{ij}^+: the ascended altitude while flying from cell i to cell j ($\Delta_{ij}^+ \geq 0$)

Δ_{ij}^-: the descended altitude while flying from cell i to cell j ($\Delta_{ij}^- \geq 0$)

The first set of constraints refers to the network flow constraints:

$$\sum_{j\in\delta_A^+} X_{Aj} = 1 \tag{3}$$

$$\sum_{i\in\delta_N^-} X_{iN} = 1 \tag{4}$$

$$\sum_{j\in\delta_i^+} X_{ij} = \sum_{j\in\delta_i^-} X_{ji} \qquad \forall i \in I | \{i \neq A \quad and \quad i \neq N\} \tag{5}$$

$$\sum_{j\in\delta_N^+} X_{Nj} = 0 \tag{6}$$

Leaving the origin and reaching the destination are guaranteed by Eq. 3 and Eq. 4, respectively. Equation 5 indicates that the drone should leave any visited cell, except for the starting cell and destination. Equation 6 assures that the path is finished just after reaching the final destination. The altitude level while arriving at each cell is adjusted using the following set of constraints.

$$h_{ij} \geqslant a_j - M(1 - X_{ij}) \qquad \forall j \in I | \{j \neq A\}, \quad \forall i \in \delta_j^- \tag{7}$$

$$h_{ij} \leqslant MX_{ij} \qquad \forall j \in I | \{j \neq A\}, \quad \forall i \in \delta_j^- \tag{8}$$

$$h_{ij} \leqslant U_j \qquad \forall j \in I | \{j \neq A\}, \quad \forall i \in \delta_j^- \tag{9}$$

Eq. 7 assures that the adjusted altitude while arriving at cell j is higher than the located obstacle (if any). If the drone does not fly from cell i to cell j, Eq. 8 sets h_{ij} equal to zero. Equation 9 defines an upper bound for the adjusted altitude while arriving at cell j. Considering the adjusted altitude, the value of altitude change for each pair of consecutive visiting cells could be calculated using the following constraints.

$$\Delta_{ij}^+ = max(h_{ij} - \sum_{g \in \delta_i^-} h_{gi}, \quad 0) \qquad \forall i \in I | \{i \neq A\}, \quad \forall j \in \delta_j^+ \tag{10}$$

$$\Delta_{ij}^- = max(\sum_{g \in \delta_i^-} h_{gi} - h_{ij} - M(1 - X_{ij}), \quad 0) \quad \forall i \in I | \{i \neq A\}, \forall j \in \delta_j^+ \tag{11}$$

$$\Delta_{Aj}^+ = max(h_{Aj} - h^A, \quad 0) \qquad \forall j \in \delta_A^+ \tag{12}$$

$$\Delta_{Aj}^- = max(h^A - h_{Aj} - M(1 - X_{Aj}), \quad 0) \qquad \forall j \in \delta_A^+ \tag{13}$$

Eq. 10–13 define the change in altitude while assuring that in case of not traveling from i to j, the related change in altitude is set to zero. This set of constraints guarantees that while flying from cell i to cell j, the altitude can be increased, decreased, or kept at the same level. The absolute change in altitude, ignoring the increasing or decreasing nature is calculated as:

$$\Delta_{ij} = \Delta_{ij}^+ + \Delta_{ij}^- \qquad \forall j \in I, \quad \forall i \in \delta_j^- \tag{14}$$

The first objective function is to minimize the path length. Considering a triangle in which the base is the direct distance between the consecutive visiting cells and the height is the absolute value of altitude change, the UAV's flying distance is calculated using the Pythagoras theorem. So, the first objective function that minimizes the aggregate flying distance is:

$$Min \quad Z_1 = \sum_{i \in I} \sum_{j \in \delta_i^+} (d_{ij}^{\ 2} + \Delta_{ij}^{\ 2}) X_{ij} \tag{15}$$

The energy consumption while traveling from cell i to cell j depends on the related fluid density of air and the flying distance. The fluid density of air is a function of flying altitude. In this research, the fluid density between cells i and j shown as ρ_{ij} is considered as the average of fluid density of air while arriving at cells i and j:

$$\rho_{ij} = (1 - 2.2558.10^{-5}(\frac{h_{ij} + \sum_{g \in \delta_i^-} h_{gi}}{2}))^{4.2577} \quad \forall i \in I | \{i \neq A\}, \forall j \in \delta_j^+ \tag{16}$$

$$\rho_{Aj} = (1 - 2.2558.10^{-5}(\frac{h_{Aj} + h^A}{2}))^{4.2577} \quad \forall j \in \delta_A^+ \tag{17}$$

Using Eq. 16 and 17, the second objective functions could be written as:

$$Min \quad Z_2 = \sum_{i \in I} \sum_{j \in \delta_i^+} \frac{\theta}{\sqrt{\rho_{ij}}} ((d_{ij}^2 + \Delta_{ij}^2) X_{ij}) + W g \Delta_{ij}^+ \tag{18}$$

To account for the maximum risk while moving from cell i to cell j the sky is decomposed into several altitude levels with different risk factors. Let's assume that a is the adjusted altitude at cell j such that $a = h_{ij}$. So, the actual altitude at cell i could be called b such that $b = h_{ij} - \Delta_{ij}^+ + \Delta_{ij}^-$. The accumulated maximum path risk could be formulated as:

$$Min \quad Z_3 = \sum_{j \in \delta_A^+} \max_{k \in [\min(a,A), \max(a,A)]} \{r_A^k\} X_{Aj} +$$

$$\sum_{i \in I | i \neq A} \sum_{j \in \delta_i^+} \max_{k \in [\min(a,b), \max(b,a)]} \{r_i^k\} X_{ij} \tag{19}$$

Finally, the fourth objective function that minimizes the number of not-visited obstacle-free cells could be formulated as follows:

$$Min \quad Z_4 = \sum_{i \in I} \sum_{j \in \delta_i^+} (1 - X_{ij})(1 - \alpha_i) \tag{20}$$

3   Solution Method

Path planning is known as an NP-hard problem [17]. Thus, these problems are generally solved using metaheuristic algorithms. Golabi *et al.* [12] reported the superiority of the NSGA-II algorithm in solving their proposed multi-objective UAV path planning model. They also showed a correlation between path length and energy consumption and tried to combine these two objective functions. As an expanded maximum-covering version of [12], this study applies a modified version of the NSGA-II algorithm to solve the proposed multi-objective model in which the first two objectives are combined to form a new objective function.

3.1   Preliminary Concepts of NSGA-II

NSGA-II is a well-known evolutionary algorithm for solving multi-objective optimization problems [7]. It starts with a randomly generated population. Each individual in the population is associated with a two-level ranking scheme. The first level refers to non-dominated sorting that assigns solutions to different Pareto fronts based on their dominance relationship. The second level, known as the crowding sort, is a strong reflection of the diversity that determines superiority/inferiority relationships between entities at the same rank based on their crowding distance. Evolution is based on a two-step process of variation based on recombination operators, and selection that results in a new generation of individuals.

3.2   The Proposed Evolutionary Components

In this study, each solution is represented as a matrix with two rows, where the first-row alleles represent the visited ground cells and the second-row alleles indicate the adjusted altitude just before entering the corresponding ground cell. It is noteworthy to mention that the first-row alleles start with the origin and end at the destination. Each first-row allele is followed by a randomly selected non-visited neighbor cell. The solution would be infeasible if all the neighbor cells are previously visited before reaching the destination. For each allele of the first row, the corresponding second-row allele would be a random integer between the first-row cell's obstacle height (if any) and the related maximum allowed flying altitude.

This study applies a one-point crossover operator that decomposes each solution into two segments. This operator generates two offspring by merging different segments of two parental chromosomes. It should be mentioned that merging the parental chromosomes from the crossover point may violate the idea of neighbor cells. In other words, the merging cells of different segments may not be neighbors. Thus, further instructions may be needed to merge the parental chromosomes and generate feasible offspring that satisfy the concept of neighbor cells. Let's assume that C is the position of the crossover point. Based on first-row genes, if the $C+1^{th}$ allele of the second parent is a neighbor cell of the Cth allele of the first parent, the first child is generated by merging all the first segment of the first parent with all the second segment of the second parent. Otherwise, until reaching a point by which merging the parents is possible, the crossover point will be iteratively shifted to the right-hand side for the selected parent or both of the parents. It could be required to add some new first and second-row genes to one of the parents before merging it with another one. This procedure is fully explained in [12]. Due to the mentioned complexity raised by changing the first-row alleles, the mutation operator is only applied to second-row genes. In this operator, a random number of second-row genes are randomly chosen and their alleles are replaced with randomly generated numbers in their allowed ranges.

The horizontal diversity of the Pareto front in multi-objective evolutionary algorithms is realized by removing extra solutions when the number of non-dominated solutions exceeds the population size [24]. NSGA-II uses the crowding distance to remove excess individuals. The main drawback of crowding distance is the lack of uniform diversity in the obtained non-dominated solutions [9]. The modified NSGA-II algorithm applied in this study uses a dynamic crowding distance method to overcome this problem [18]. The main idea of this method is to remove the individual with the lowest crowding distance value, followed by the recalculation of crowding distance for the remaining solutions.

The modified NSGA-II algorithm also benefits from a rank-based roulette wheel selection to choose better solutions for recombination [1]. This operator uses crowding distance to rank the solutions in the non-dominated set. The solution with a better rank would have a higher chance of being selected as a potential parent in the mating pool. Last but not least, the developed modified

NSGA-II applies a local search operator on some randomly selected obstacle-free second-row genes of offspring generated through the mutation operation by copying the preceding allele. This operator refrains from unnecessary changes in altitude levels.

4 Results and Discussion

The modified NSGA-II algorithm is applied to solve the developed mathematical model on a partial map of Berlin city imported using 3D City Database Importer/Exporter[1] and FME Data Inspector[2] to extract the Geography Markup Language (.gml) files. The map is used as a large scenario based on the number of considered buildings treated as static obstacles, and based on different origins and destinations (see Fig. 2). Polygon triangulation is used to partition buildings with different polygon shapes into a set of triangles with pairwise non-intersecting interiors. Each scenario is represented as a grid-based map with equal-size cells, such that each cell is identified with a unique index. A cell is considered obstacle-free if it is not occupied by any triangle of considered buildings.

To evaluate the performance of the modified NSGA-II, the same scenario is also solved using a classic NSGA-II algorithm. The comparisons are based on 50,000 function evaluations, with 15 independent runs on each scenario. For each algorithm, the applied parameters comprising the population size, the crossover probability, the mutation probability, the mutation rate, and selection pressure are tuned using MAC [23] as an automated algorithm configuration tool for multi-objective optimization. The experiments in MAC are executed subject to 100 function evaluations. For each generated configuration, experiments are

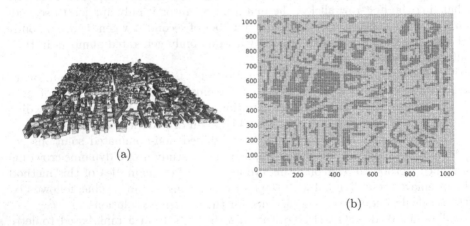

(a)

(b)

Fig. 2. An example of a considered scenario.

[1] https://www.3dcitydb.org/3dcitydb/3dimpexp/.
[2] https://www.safe.com/transformers/inspector/.

repeated 5 times to reduce the noise for the adopted surrogate model in MAC. Fine-tuned algorithms are implemented to solve the generated scenario extracted from the map of Berlin city. The effect of using MAC on improving the obtained hypervolume is illustrated in Fig. 3.a. Furthermore, Fig. 3.b exemplifies the non-dominated solutions of one of the considered scenarios obtained from the modified NSGA-II algorithm. Generally, the obtained results indicate the superiority of the modified NSGA-II algorithm based on the hypervolume metric, as well as the obtained number of non-dominated solutions.

The results obtained from implementing the solution algorithm on the considered occupancy grid map highlight the effect of the new objective function added to minimize the number of non-visited obstacle-free cells. The algorithm is able to obtain longer paths due to the trade-off between conflicting objective functions. Besides, the continuous variable for flying altitude provides more realistic solutions. It is noteworthy to mention that the algorithm is coded in MATLAB R2022a and implemented on an Intel Core i7-6500U CPU @ 2.50 GHz laptop with 8 GB RAM, 6 MB L3 Cache, and 1 MB L2 Cache. It took almost 3200 s seconds to solve the problem on a grid map of size 10000 cells.

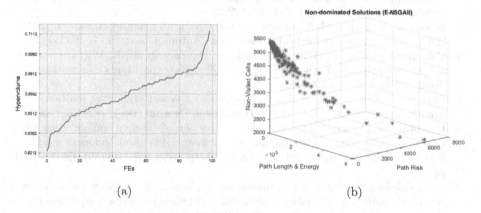

(a)   (b)

Fig. 3. Examples of: (a) convergence plot of hypervolume; (b) non-dominated solutions obtained using modified NSGA-II.

5   Conclusion

Considering both the possibilities of bypassing obstacles or adjusting the altitude to fly above them, this work studies a multi-objective offline maximum-covering path planning problem. Considering a continuous variable altitude level, this study aims to minimize the flying distance, the required energy, the maximum path risk, and the number of not-visited obstacle-free cells, simultaneously. The studied problem is formulated by a novel mathematical model. The applicability of the developed model is checked using real scenarios generated from a partial map of Berlin city. Using a rank-based roulette wheel selection and a dynamic crowding distance method, the problem is solved by a modified NSGA-II algorithm hybridized with a local search operator. Considering a dynamic online environment consisting of both static and moving obstacles could be an interesting guideline for future studies.

References

1. Al Jadaan, L.O., Rao, C., Rajamani, L.: Ranked based roulette wheel selection method. In: International Symposium on Recent Advances in Mathematics and its Applications:(ISRAMA 2005), Calcutta Mathematical Society at AE-374, Sector-1, Salt Lake City Kolkata (Calcutta), vol. 700064 (2005)
2. Ali, Z.A., Zhangang, H., Hang, W.B.: Cooperative path planning of multiple UAVs by using max-min ant colony optimization along with Cauchy mutant operator. Fluct. Noise Lett. **20**(01), 2150002 (2021)
3. Alotaibi, E.T., Alqefari, S.S., Koubaa, A.: LSAR: Multi-UAV collaboration for search and rescue missions. IEEE Access **7**, 55817–55832 (2019)
4. Basiri, A., Mariani, V., Silano, G., Aatif, M., Iannelli, L., Glielmo, L.: A survey on the application of path-planning algorithms for multi-rotor UAVs in precision agriculture. J. Navig. **75**(2), 364–383 (2022)
5. Bolourian, N., Hammad, A.: Lidar-equipped UAV path planning considering potential locations of defects for bridge inspection. Autom. Constr. **117**, 103250 (2020)
6. Calcara, A., Gilli, A., Gilli, M., Marchetti, R., Zaccagnini, I.: Why drones have not revolutionized war: the enduring hider-finder competition in air warfare. Int. Secur. **46**(4), 130–171 (2022)
7. Deb, K., Pratap, A., Agarwal, S., Meyarivan, T.: A fast and elitist multiobjective genetic algorithm: NSGA-II. IEEE Trans. Evol. Comput. **6**(2), 182–197 (2002)
8. Debnath, S.K., Omar, R., Latip, N.A.: Comparison of different configuration space representations for path planning under combinatorial method. Indonesian J. Elect. Eng. Comput. Sci. **1**(1), 401–408 (2019). http://doi.org/10.11591/ijeecs.v14.i1.pp1-8
9. Dhanalakshmi, S., Kannan, S., Mahadevan, K., Baskar, S.: Application of modified NSGA-II algorithm to combined economic and emission dispatch problem. Int. J. Elect. Power Energy Syst. **33**(4), 992–1002 (2011)
10. Dorling, K., Heinrichs, J., Messier, G.G., Magierowski, S.: Vehicle routing problems for drone delivery. IEEE Trans. Syst. Man Cybern. Syst. **47**(1), 70–85 (2016)
11. Ghambari, S., Lepagnot, J., Jourdan, L., Idoumghar, L.: UAV path planning in the presence of static and dynamic obstacles. In: 2020 IEEE Symposium Series on Computational Intelligence (SSCI), pp. 465–472. IEEE (2020)
12. Golabi, M., Ghambari, S., Lepagnot, J., Jourdan, L., Brévilliers, M., Idoumghar, L.: Bypassing or flying above the obstacles? A novel multi-objective UAV path planning problem. In: 2020 IEEE Congress on Evolutionary Computation (CEC), pp. 1–8. IEEE (2020)
13. Golabi, M., Shavarani, S.M., Izbirak, G.: An edge-based stochastic facility location problem in UAV-supported humanitarian relief logistics: a case study of Tehran earthquake. Nat. Hazards **87**(3), 1545–1565 (2017)
14. González, D., Pérez, J., Milanés, V., Nashashibi, F.: A review of motion planning techniques for automated vehicles. IEEE Trans. Intell. Transp. Syst. **17**(4), 1135–1145 (2015). https://doi.org/10.1109/TITS.2015.2498841

15. Jing, W., Deng, D., Wu, Y., Shimada, K.: Multi-UAV coverage path planning for the inspection of large and complex structures. arXiv preprint arXiv:2007.13065 (2020)
16. Konatowski, S., Pawłowski, P.: Application of the ACO algorithm for UAV path planning. Przeglad Elektrotechniczny **95**(7), 115–118 (2019). https://doi.org/10.15199/48.2019.07.24
17. Li, G., Yamashita, A., Asama, H., Tamura, Y.: An efficient improved artificial potential field based regression search method for robot path planning. In: 2012 IEEE International Conference on Mechatronics and Automation, pp. 1227–1232. IEEE (2012)
18. Luo, B., Zheng, J., Xie, J., Wu, J.: Dynamic crowding distance? A new diversity maintenance strategy for MOEAs. In: 2008 Fourth International Conference on Natural Computation, vol. 1, pp. 580–585. IEEE (2008)
19. Motlagh, N.H., Bagaa, M., Taleb, T.: UAV-based IoT platform: a crowd surveillance use case. IEEE Commun. Mag. **55**(2), 128–134 (2017)
20. Paredes, J.A., Saito, C., Abarca, M., Cuellar, F.: Study of effects of high-altitude environments on multicopter and fixed-wing UAVs' energy consumption and flight time. In: 2017 13th IEEE Conference on Automation Science and Engineering (CASE), pp. 1645–1650. IEEE (2017)
21. Poikonen, S., Golden, B.: Multi-visit drone routing problem. Comput. Oper. Res. **113**, 104802 (2020)
22. Popović, M., Hitz, G., Nieto, J., Sa, I., Siegwart, R., Galceran, E.: Online informative path planning for active classification using UAVs. In: 2017 IEEE International Conference on Robotics and Automation (ICRA), pp. 5753–5758. IEEE (2017)
23. Rakhshani, H., Idoumghar, L., Lepagnot, J., Brévilliers, M.: MAC: many-objective automatic algorithm configuration. In: Deb, K., et al. (eds.) EMO 2019. LNCS, vol. 11411, pp. 241–253. Springer, Cham (2019). https://doi.org/10.1007/978-3-030-12598-1_20
24. Ramesh, S., Kannan, S., Baskar, S.: Application of modified NSGA-II algorithm to multi-objective reactive power planning. Appl. Soft Comput. **12**(2), 741–753 (2012)
25. Shavarani, S.M., Golabi, M., Izbirak, G.: A capacitated biobjective location problem with uniformly distributed demands in the UAV-supported delivery operation. Int. Trans. Oper. Res. **28**(6), 3220–3243 (2021)
26. Shavarani, S.M., Mosallaeipour, S., Golabi, M., İzbirak, G.: A congested capacitated multi-level fuzzy facility location problem: an efficient drone delivery system. Comput. Oper. Res. **108**, 57–68 (2019)
27. Shivgan, R., Dong, Z.: Energy-efficient drone coverage path planning using genetic algorithm. In: 2020 IEEE 21st International Conference on High Performance Switching and Routing (HPSR), pp. 1–6. IEEE (2020). https://doi.org/10.1109/HPSR48589.2020.9098989
28. Song, B., Qi, G., Xu, L.: A survey of three-dimensional flight path planning for unmanned aerial vehicle. In: 2019 Chinese Control And Decision Conference (CCDC), pp. 5010–5015. IEEE (2019). https://doi.org/10.1109/CCDC.2019.8832890
29. Wyard, C., Beaumont, B., Grippa, T., Hallot, E.: UAV-based landfill land cover mapping: optimizing data acquisition and open-source processing protocols. Drones **6**(5), 123 (2022)
30. Yang, L., Qi, J., Xiao, J., Yong, X.: A literature review of UAV 3D path planning. In: 2014 11th World Congress on Intelligent Control and Automation (WCICA), pp. 2376–2381. IEEE (2014). https://doi.org/10.1109/WCICA.2014.7053093

31. Yin, C., Xiao, Z., Cao, X., Xi, X., Yang, P., Wu, D.: Offline and online search: UAV multiobjective path planning under dynamic urban environment. IEEE Internet Things J. **5**(2), 546–558 (2017). https://doi.org/10.1109/JIOT.2017.2717078

32. Yu, X., Li, C., Zhou, J.: A constrained differential evolution algorithm to solve UAV path planning in disaster scenarios. Knowl. Based Syst. **204**, 106209 (2020). https://doi.org/10.1016/j.knosys.2020.106209

33. Zhao, Y., Zheng, Z., Liu, Y.: Survey on computational-intelligence-based UAV path planning. Knowl. Based Syst. (2018). https://doi.org/10.1016/j.knosys.2018.05.033

34. Zhou, X., Yi, Z., Liu, Y., Huang, K., Huang, H.: Survey on path and view planning for UAVs. Virtual Real. Intell. Hardw. **2**(1), 56–69 (2020). https://doi.org/10.1016/j.vrih.2019.12.004

An Elitist Non-dominated Heuristic Resolution for the Dynamic Asset Protection Problem

Quentin Peña[✉], Aziz Moukrim, and Mehdi Serairi

Université de Technologie de Compiègne, CNRS, Heudiasyc (Heuristics and Diagnosis of Complex Systems), CS, 60 319 - 60 203 Compiègne Cedex, France
{quentin.pena,aziz.moukrim,mehdi.serairi}@hds.utc.fr

Abstract. During escaped wildfires, community assets are at risk of damage or destruction. Preventive operations requiring dispatching resources and cooperation can be taken to protect these assets. The planning of such operations is sensitive to unforeseen disruptions that may occur. To account for the effects of the disruption, it may be necessary to alter the initial routes of the vehicles. The problem rising from the rescheduling of the vehicles is a bi-objective optimization problem known as the Dynamic Asset Protection Problem (D-APP). We propose a genetic algorithm based on the Non-dominated Sorting Genetic Algorithm (NSGA-II) to solve the D-APP. We define new mutation and crossover operators adapted to our problem, and we propose procedures to repair and evaluate a solution based on Mixed Integer Programming (MIP).

Keywords: bi-objective optimization · vehicle routing · team orienteering · synchronization · NSGA-II

1 Introduction

In the recent years, wildfires break out more frequently throughout the world. When wildfires are not controlled, they quickly expand and can burn thousands of hectares of vegetation. In urban areas, the fire can also harm people and damage infrastructure. Emergency response teams and resources must be deployed to respond to these escaped wildfires. Multiple operations are jointly carried out, from fire containment to evacuation, sheltering operations and including asset protection. In this paper, we will focus on the preventive actions for the protection of community assets.

Depending on the community asset, different actions can be taken to mitigate or nullify the damages caused when the wildfire reach them. Such actions include removing fuel materials, wetting down buildings, or reducing fire. Preventive protection actions must be taken in a timely manner: it has to be performed before the fire reaches the asset, but not too early to be efficient. Some interventions may require several trucks with specific capacities, thus requiring different teams to collaborate to perform the task in a synchronous way. In particular, we will focus on rerouting the vehicles after a disruption occurs that invalidates their initial routes. A wide range of disruptions can impact our initial plans in different ways. For example, we may not have all the resources available, due to a faulty equipment or a vehicle breakdown. The time windows on some assets may be updated after unforeseen wind or weather changes, altering the propagation of the fire. Travel times between assets may also change if traffic jams are caused by people evacuating, or a road might be blocked by a fallen tree.

The problem of routing vehicles for preventive protection operations can be viewed as a variant of the Team Orienteering Problem (TOP) with time windows and synchronization constraints. This problem was first introduced by Merwe et al. [9] as the Asset Protection Problem (APP). The authors proposed a Mixed Integer Linear Programming (MILP) formulation of the problem. Later, Merwe et al. [10] introduced the dynamic APP (D-APP), which is a bi-objective problem for rerouting the vehicles after a disruption. The authors updated the MILP formulation from the mono-objective version of the problem to account for the deviation. They generated theset of solutions offering optimal trade-off between protection of the assets and deviation from the initial routes using an ϵ-constraint scheme. Peña et al. [13] proposed a new mathematical formulation and valid inequalities based on the properties of the D-APP.

In this paper, we will present a heuristic solution method for the D-APP based on the Non-dominating Sorting Genetic Algorithm (NSGA-II) [3]. We will introduce different crossover and mutation operators specific to our problem, including a destruction/construction operator as well as different MILP to repair and evaluate a solution.

2   Dynamic Asset Protection Problem

During a wildfire, community assets such as schools, hospitals, bridges are at risk of being damaged. A fleet of heterogeneous vehicles must be dispatched to the different assets to perform preventive protection operations. These operations must be accomplished within a specific time window, and often require the cooperation of multiple vehicles.

An asset is protected if, within its time window, enough vehicles are present at the asset to accomplish the protection operation. The protection of an asset requires some resources (e.g., crew size, number of fire hoses, ...), that need to be met by the vehicles assigned to the asset.

In the dynamic APP, we already have routes assigned to the vehicles. However, an unforeseen disruption occurred and these initial routes may no longer be feasible nor optimal. We want to recompute the routes to take into account the consequences of the disruption. We then have two competing objectives:

- maximizing the total value of the protected assets
- minimizing the deviation from initial routes

We define the deviation as the number of vehicle/asset reassignments, i.e., if an asset is added to or removed from the initial route of a vehicle, then it implies a deviation of one.

2.1 Problem Presentation

An instance of our problem is represented by a graph $G = (V, A)$, with V representing the locations and A the arcs. There are n total locations. The first m locations represent the depots from which the vehicles depart, and the n-th location is a fictitious sink node. The remaining $n - m - 1$ locations represent the assets to protect. We define subsets of V: the depots V^d and the assets V^a. Each asset i has a value v_i, a requirement vector r_i, a service duration a_i and a time window $[o_i; c_i]$. The set $\mathcal{P}$ represents the available vehicles. Each vehicle p has a capability vector cap_p. In order to be protected, the vehicles assigned to an asset must collectively meet the resource requirement of the asset. For example, an asset with resource requirement $r_i = (1, 2, 1)$ can be protected by vehicles p and q with respective capability $cap_p = (1, 1, 0)$ and $cap_q = (1, 1, 1)$. All the vehicles assigned to the asset must be present when the service starts, and throughout the entirety of the service. Parameters t_{ijp} are the travel time between locations i and j for vehicle p. Travel times satisfy the triangle inequality. We note Φ the solution representing the routes of the vehicles before disruption.

Before proceeding further, we introduce some definitions. An arc between two assets i and j is called a *valid arc* for vehicle p if $o_i + a_i + t_{ijp} \leq c_j$. In other words, vehicle p can visit asset i before asset j within the respective time windows of the assets. Additionally, we say that the insertion of an asset k between two assets i and j in the route of vehicle p is at a valid position if arcs (i, k) and (k, j) are valid arcs for vehicle p.

2.2 Bi-Objective Optimization

The D-APP is a bi-objective optimization problem. We recall some terminology related to Multi-objective Optimization Problems (MOP).

In MOP, a solution is evaluated according to an objective function vector $f = (f_1, ..., f_d)$ with d objectives. Without loss of generality, we suppose that all the objectives are to be maximized. These d objectives are competing against each other: improving one of the objective will often degrade one or multiple other objectives. Hence, we want to find the set of efficient solutions based on a dominance relation between solutions [7].

Definition 1. *Let u and v be vectors of $\mathbb{R}^d$, we say that u dominates v if and only if $u_i \geq v_i$ for each $i \in \{1, ..., d\}$ and there exists $j \in \{1, ..., d\}$ such that $u_j > v_j$. We denote this dominance relation by $u \succ v$.*

Definition 2. *A solution s is efficient if there is no other solution s' such that $f(s') \succ f(s)$, with $f(s)$ the objective function vector associated with solution s.*

For ease of use, we say that a solution s dominates a solution s' if and only if its objective function vector $f(s)$ dominates $f(s')$. The set of all efficient solutions is known as the efficient set. The set of objective vectors with respect to the efficient set is call the non-dominated set, or Pareto front [12].

In many vehicle routing problems, multiple competing objectives have been considered [6]. A popular approach is to solve the multi-objective problem using a decomposition approach. The multi-objective problem is decomposed in multiple single objective problems, using aggregation functions. For instance, the bi-objective traveling salesman problem has been solved using ant colony optimization based on decomposition [2]. A metaheuristic method that combines a Pareto ant colony optimization algorithm and a variable neighborhood search method has been proposed for the bi-objective TOP (BTOP) [14]. Finally, a two-phase decomposition method based on Local Search has been proposed to solve Selective Pickup and Delivery Problems with Time Windows (SPDPTW) [1].

Several approaches extend the fast and elitist Non-Dominated Sorting Genetic Algorithm (NSGA-II) [3]. It has been efficiently applied to various multi-objective problems including but not limited to the BTOP [11], the Green Vehicle Routing Problem [4] and the Vehicle Routing Problem with Route Balancing [5].

3   NSGA-II

In this section, we will discuss the implementation of the NSGA-II algorithm to solve the D-APP. We will first present in Sect. 3.1 an overview of the NSGA-II algorithm. We will then present in Sect. 3.2 the encoding we encounter in the literature for a genetic algorithm on a problem similar to the problem at hand. We will introduce mutation and crossover operators based on the properties of our problem in Sect. 3.3. Finally, we will define two different procedures for repairing and evaluating a solution in Sect. 3.4.

3.1   Overview

NSGA-II is an iterative algorithm. For each generation t, we consider a population R_t of size $2N$, that is the combination of two subpopulations of size N: P_t, the parents, and Q_t, the offspring. There are three main steps in the NSGA-II algorithm, described below. A solution i has two fitness criteria relative to the current population: a rank r_i and a crowding distance d_i. The rank represents the quality of the solution with regards to the dominance relation presented in Sect. 2.2. The crowding distance represents the quality of the solution in terms of diversification. For more information on how these criteria are computed, we refer the reader to [3].

At generation t, the three steps are:

Step 1 - Initialization. Create the population R_t by combining the parent and offspring populations. Compute the rank of the solutions in R_t and identify all the non-dominated fronts $\mathcal{F} = (\mathcal{F}_1, \mathcal{F}_2, ...)$. Compute the crowding distance of the solutions within each non-dominated front.

Step 2 - Parent population selection. Create the parent population for next generation P_{t+1} by selecting the N solutions from population R_t. Between two solutions with different ranks, we prefer the solution with the lowest rank. If both solutions belong to the same front, we prefer the solution with the lowest crowding distance.

Step 3 - Offspring creation. Create offspring population Q_{t+1} from P_{t+1}. Details are given in Algorithm 1. The tournament operator is binary tournament, as described in [3]. Two solutions are selected at random, the solution with lowest rank is selected, or with lowest crowding distance if there is a tie. The crossover and mutation operators are discussed in Sect. 3.3. The repair and evaluation procedure is discussed in Sect. 3.4.

Algorithm 1. Offspring creation

Data: Parent population P, mutation rate μ
Result: Offspring population Q
1: $Q \leftarrow \emptyset$;
2: **while** $|Q| \leq N$ **do**
3:     $p_1 \leftarrow tournament(P)$;
4:     $p_2 \leftarrow tournament(P)$;
5:     $s \leftarrow crossover(p_1, p_2)$; (See Section 3.3)
6:     **if** rand() $< \mu$ **then**
7:         $s \leftarrow mutate(s)$; (See Section 3.3)
8:     **end if**
9:     $s \leftarrow repair\_and\_evaluate(s)$; (See Section 3.4)
10:    $Q \leftarrow Q \cup \{s\}$;
11: **end while**
12: **return** Q

3.2   Encoding

We based the implementation of the NSGA-II algorithm for our problem on a genetic algorithm proposed for the mono-objective version of the APP with a homogeneous fleet of vehicles [8].

A solution s is represented by an array of integers, representing the order in which assets are visited for each vehicle. The route of a vehicle always starts at a depot and ends at the sink node. For instance, there are three vehicles in solution $[1, 2, 6, 4, 11, 1, 5, 7, 3, 11, 1, 7, 3, 11]$, the route of the first vehicle is $(1 \rightarrow 2 \rightarrow 6 \rightarrow 4 \rightarrow 11)$, the second $(1 \rightarrow 5 \rightarrow 7 \rightarrow 3 \rightarrow 11)$ and the last $(1 \rightarrow 7 \rightarrow 3 \rightarrow 11)$.

We note $\mathcal{P}_i^s$ the set of vehicles assigned to asset i in solution s, and $\overline{\mathcal{P}_i^s}$ the set of available vehicles not assigned to asset i in solution s.

3.3   Operators

Valid Crossover Operator (CXVAL). This crossover operator between two solutions s_1 and s_2 selects a vehicle at random. The route for this vehicle in

s_1 is cut after a random asset i_k. The route for this vehicle in s_2 is also cut, after asset j_l. The offspring route for this vehicle is constructed by the taking the part of the route up to, and including, asset i_k in s_1 first, and then the part of the route after asset j_l in s_2. For example, suppose we have two routes $(i_1 \rightarrow i_2 \rightarrow \mathbf{i_3} \rightarrow i_4 \rightarrow i_5)$ and $(j_1 \rightarrow j_2 \rightarrow j_3 \rightarrow \mathbf{j_4} \rightarrow j_5 \rightarrow j_6)$, and assume the cuts happen after assets i_3 and j_4 respectively, indicated in bold. The resulting route would be $(i_1 \rightarrow i_2 \rightarrow i_3 \rightarrow j_5 \rightarrow j_6)$. The route of the second vehicle is cut in a way such that arc (i_k, j_{l+1}) is a valid arc. This crossover may result in duplicate assets in the route of a vehicle; we only keep the first occurrence of an asset to fix this issue.

Time Crossover Operator (CXTIM). This crossover operator between two solutions s_1 and s_2 selects a time at random within the time horizon. The routes for the vehicles in s_1 are cut when the start time of service of the asset exceeds the chosen time, and represent the first part of the offspring routes. We then cut the routes of the vehicles in s_2 such that there is a valid arc between the last asset of the first part of the route and the first asset of the second part of the route.

Single-Change Operators. We define two different mutation operators that perform a single change on the solution, with same probability of being used: an insertion operator and a removal operator.

Insertion Operator. The insertion operator adds one randomly selected asset to the route of one or multiple vehicles. An asset is selected at random. The asset is added at a random valid position in the route of vehicles, taken in a random order, until the resource requirement of the asset is met.

Removal Operator. The removal operator removes one randomly selected asset from the route of one or multiple vehicles. An asset is selected at random. The asset is removed from the route of all the vehicles.

Multi-change Operator. We define a mutation operator that performs multiple changes on the solution, first removing multiple assets from the solution in the destruction phase, then inserting multiple protected assets in the construction phase.

During the destruction phase, the operator randomly selects d assets to be removed from the current solution. The number of assets removed is randomly selected between 1 and d_{max}. The destruction parameter d_{max} is initially set to 3. If there is no improvement on the optimal Pareto front $\mathcal{F}_1$, its value is increased, and resets to 3 when an improvement is found. In the random selection process, we can assign weights to the assets in order to favor removing assets that induce most deviation. We note w_i^- the weight associated to asset i. The probability of selecting asset i to be removed is thus $p^-(i) = w_i^- / \sum_i w_i^-$. If $w_i^- = 1$ for all assets, we have fully random behavior. Alternatively, we can use a weight based on the deviation induced by the removal of asset i from solution s, with γ a parameter to be determined:

$$w_i^- = \left(1 + max(0, |\mathcal{P}_i^s| - |\overline{\mathcal{P}_i^s}|)\right)^\gamma \tag{1}$$

During the construction phase, the operator uses a Best Insertion Heuristic (BIH) to insert a subset of assets to the current solution. The number of assets to add is chosen randomly between d and $d + c_{max}$. The construction parameter c_{max} is initially set to 3. The assets to be inserted are randomly selected. We can assign weights to the assets in the selection process. We note w_i^+ the weight associated to asset i. We can use a weight based on the profit v_i associated with the protection of asset i and a lower bound on the deviation necessary for the protection of the asset nb_i^+, with α and β parameters to be defined:

$$w_i^+ = v_i^\alpha / (1 + nb_i^+)^\beta \tag{2}$$

We want to add each asset to the route of enough vehicles for the resource protection to be met. We also want to minimize the number of vehicles we use to protect the asset. As we do not know how many vehicles will be required to meet the resource requirement, we will generate multiple insertion patterns and select the one minimizing our criterion. We detail the process in Algorithm 2. In order to account for the deviation from the pre-disruption routes, we first select the vehicles for which the asset is in the pre-disruption route. If these vehicles are not sufficient to meet the resource requirement, we continue the process with the remaining vehicles. We select the vehicles in a random order, until the protection requirement is met.

Adaptive Parameters. The multi-change operator relies on parameters α, β and γ to control the relative importance of the different factors when associating weights to assets. They are first initialized with $\alpha = 1$, $\beta = 1$ and $\gamma = 0.5$, and then adaptively tuned during the offspring creation phase. We generate M offspring solutions with slightly different values of α, β and γ. The values leading to the best offspring subpopulation are recorded to be used in the next iteration. All the offspring solutions generated are considered in the offspring population Q of the current step.

3.4 Repair and Evaluation Procedure

A solution is represented by the route of each vehicle. It is sufficient to know the routes of the vehicles to compute the deviation from the pre-disruption routes. However, we cannot determine which assets are effectively protected: we must check if it is possible to synchronize the visits of all assigned vehicles within the time window of the asset, and if the resource requirement is met by these vehicles.

Some solutions are not feasible. For instance, two vehicles may visit two assets in a different order, thus causing the synchronization to be impossible.

Algorithm 2. Construction: Add an asset

Data: Solution S, asset k, number of insertion patterns nb_p
Result: A solution that protects asset k, if possible
1: **if** the available vehicles cannot meet the resource requirement **then**
2:    **return** S
3: **end if**
4: **for** $cpt = 1...nb\_p$ **do**
5:    $V_{cpt} \leftarrow \emptyset$; {Set of selected vehicles at iteration cpt}
6:    $cost_{cpt} = 0$;
7:    Determine a random order on the vehicles that prioritizes vehicles in $\mathcal{P}_k^\Phi$
8:    **for** each vehicle p following the previously defined order **do**
9:        **if** there is a valid position in the route of vehicle p **then**
10:            $V_{cpt} \leftarrow V_{cpt} \bigcup \{p\}$
11:            $cost_{cpt} \leftarrow cost_{cpt} + 1$
12:            **if** vehicles in V_{cpt} meet the resource requirement of asset k **then**
13:                Begin new insertion pattern (next cpt)
14:            **end if**
15:        **end if**
16:    **end for**
17: **end for**
18: Select set of vehicles V_* with lowest cost
19: Insert asset k in the routes of vehicles in V_* in solution S
20: **return** S

Our repair procedure aims at finding the best subroutes of the solution, to make it feasible and maximize total protected value. We do not modify the order in which assets are visited by a vehicle, nor do we add new assets to the routes. The repair procedure determines which assets can actually be protected, thus contributing to the total protected value. It also gives data to correct the deviation, if unprotected assets have been added to the route of a vehicle for instance. At the end of the repair procedure, we know the value of the two objective functions for the solution we have just repaired. Hence, we can use the repair procedure as the evaluation procedure for our solutions. By doing so, we also ensure that all the solutions we consider are feasible.

We propose two different MIPs used for repairing and evaluating solutions for our problem. We note $\mathcal{P}_i$ the set of vehicles that have asset i in their route. We note X_p the set of arcs (i_k, i_l) between assets in the route of vehicle p, with $k < l$.

Asset Penalization. The first MIP tries to find a feasible solution from the given routes. Assets can be visited outside of their time windows, but these assets cannot be protected. Infeasibilities are lifted by removing assets entirely from the solution.

We define three sets of decision variables:

- Binary variables Y_i, set to 1 if asset i is protected. Asset i is protected when service starts within its time window and its resource requirement is met.

- Binary variables θ_i, set to 1 if asset i is removed from the solution.
- Continuous variables S_i, that represent the start time of service of asset i.

$$Maximize \sum_i v_i Y_i \tag{3}$$

$$(1 - \theta_i) \sum_p cap_p \geq r_i Y_i \ \forall i \in V^a \tag{4}$$

$$S_i + t_{ijp} + a_i \leq S_j + M_1(\theta_i + \theta_j) \ \forall p \in \mathcal{P}, (i,j) \in X_p \tag{5}$$

$$o_i - M_2(1 - Y_i) \leq S_i \leq c_i + M_2(1 - Y_i) \ \forall i \in V^a \tag{6}$$

$$Y_i \in \{0,1\}, \ \theta_i \in \{0,1\}, \ S_i \in \mathbf{R} \ \forall i \in V^a \tag{7}$$

Objective function (3) maximizes the total protected value.

Constraints (4) ensure that the protection requirement is met for protected assets. Assets that have been removed from the solution (with $\theta_i = 1$) cannot be protected. Constraints (5) set correct start time of service for assets i and j when asset i is visited by the vehicle before asset j. The order in which the assets are visited is fixed within the solution. However, as assets can be removed, we need to consider every pair of assets (i,j) visited by the vehicle such that asset i is visited before asset j. Constraints (6) ensure that a protected asset is visited within its time window. Constraints (7) define the domain of the decision variables.

Assignment Penalization. The second MIP tries to find a feasible solution from the given routes. Infeasibilities are lifted by removing assets from the routes of individual vehicles.

We use binary variables Y_i and continuous variables S_i. We replace variables θ_i by variables θ_{pi}, set to 1 if asset i is removed from the route of vehicle p.

$$Maximize \sum_i v_i Y_i \tag{8}$$

$$\sum_{p \in \mathcal{P}_i} (1 - \theta_{pi}) cap_p \geq r_i Y_i \ \forall i \in V^a \tag{9}$$

$$S_i + t_{ijp} + a_i \leq S_j + M_1(\theta_{pi} + \theta_{pj}) \ \forall p \in \mathcal{P}, (i,j) \in X_p \tag{10}$$

$$Y_i + \theta_{pi} \geq 1 \ \forall p, \ \forall i \in V^a \tag{11}$$

$$o_i - M_2(1 - Y_i) \leq S_i \leq c_i + M_2(1 - Y_i) \ \forall i \in V^a \tag{12}$$

$$Y_i \in \{0,1\}, \ S_i \in \mathbb{R} \ \forall i \in V^a \tag{13}$$

$$\theta_{pi} \in \{0,1\} \ \forall i \in V^a, \ p \in \mathcal{P}_i \tag{14}$$

Objective function (8) maximizes the total protected value.

Constraints (9) ensure that the protection requirement is met for protected assets. If asset i is removed from the route of the vehicle (with $\theta_{pi} = 1$), the vehicle does not contribute to the protection. Constraints (10) set correct start time of service for assets i and j when asset i is visited by the vehicle before asset j, similarly to constraints (5). Constraints (11) ensure that unprotected assets are removed from the routes of all vehicles. Constraints (12) ensure that a protected asset is visited within its time window. Constraints (13) and (14) define the domain of the decision variables.

Local Search. After repairing a solution, we explore its neighborhood to find a dominating solution. We base our local search on the MIP used in the ϵ-constraint method for the D-APP introduced in [13]. We use the MIP that maximizes total protected value with deviation limited to the value of the deviation of the solution we are considering. This solution is used as a warm-start for the MIP. We set a high relative gap tolerance in our solver, meaning that the resolution will stop before optimality is proven. For example, with a tolerance of 0.05, a solution is returned when it is proved to be within 5% of optimal.

4   Computational Results

We carried out computational testing on a computer with an Intel Core i7-8550U processor and 8 GB of RAM. We implemented the method in Julia.

We generated 10 benchmark instances[1], following the guidelines provided by Merwe et al. [9]. Each instance has 100 assets randomly distributed within a 80 km by 80 km grid. Instances of less than 100 assets are created using a subset of the 100-asset instances.

In order to evaluate our algorithm performance, we will use a quality indicator to compare approximate PFs: the hypervolume (HV) [15]. The hypervolume (or S-volume) is widely used in multi-objective optimization as we can compute it without knowing the optimal PF. We suppose, without loss of generality, that we want to maximize objective function f_1 and minimize objective function f_2. The hypervolume requires two reference points in order to be computed: it is important to use the same reference points when we compare two approximate fronts. For a set of approximate fronts, the references points called nadir and ideal are defined as $nadir = (f_1^{min}, f_2^{max})$ and $ideal = (f_1^{max}, f_2^{min})$, where f_i^{min} and f_i^{max}, $i = 1, 2$ refer to the minimum and maximum values of objective functions f_1 and f_2 encountered in the set of approximate fronts. Let $\Lambda(a_i)$ be the size of the rectangular area a_i constructed with a solution s_i from an approximation set A and the nadir as corners. For approximate set A, we compute the HV as follows:

$$HV(A) = \frac{\Lambda\left(\bigcup_{a_i \in A} a_i\right)}{(f_1^{max} - f_1^{min})(f_2^{max} - f_2^{min})} \tag{15}$$

---

[1] See https://www.hds.utc.fr/~penaquen for the detailed instances and pre-disruptions routes.

4.1 Mutation Rate Tuning

In this section, we want to test the influence of the mutation rate μ on the output of our algorithm. We launched the algorithm five times with a time limit of 60 s on five training instances with 30, 40, 50 and 60 assets, with two different vehicle breakdowns as the disruption.

Based on preliminary tuning work, we used fixed values for some of our parameters. The population size is set to $N = 100$. We use the time crossover operator as crossover operator and multi-change operator as mutation operator. For the choice criteria w_i^- and w_i^+, the parameters are set to $\alpha = 1.0$, $\beta = 0.5$ and $\gamma = 1.0$. Destruction and construction parameters c_{max} and d_{max} are initially set to 3. The initial population is generated by applying the multi-change operator with high c_{max} and d_{max} values on the solution representing the initial routes.

We report in Fig. 1 the average gap between the hypervolume of the non-dominated front $\mathcal{F}_1$ obtained with each mutation rate and the hypervolume of the best known Pareto front.

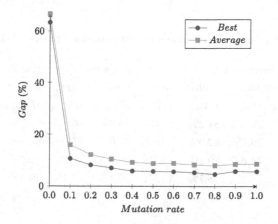

Fig. 1. Average gap between the front obtained by NSGA-II and best known front, based on mutation rate μ.

We can see that our mutation operator impacted the quality of the front we generated. We obtained the worst results when the mutation operator was disabled ($\mu = 0$), with a gap superior to 63%. The gap steeply decreased to 15% on average when the mutation operator was enabled and steadily decreased, dropping below 9% on average for $\mu = 0.5$. The gap reached its lowest point for $\mu = 0.8$ and slightly deteriorated for higer mutation rates.

4.2 Performance Analysis

In this section, we test the influence of our evaluation models and operators, and the impact of our additional components. Following preliminary work, we chose not to consider the valid crossover operator (CXVAL). Hence, we will only present results using the time crossover operator (CXTIM). We will first compare the results obtained using our two different evaluation models, with our single-change operators and our multi-change operator. Then, we will evaluate the impact of the adaptive scheme and the local search procedure we presented.

We launched our NSGA-II algorithm five times with the different sets of operators for each of our ten benchmark instances with 30, 40, 50 and 60 assets, and two different random vehicle breakdowns as the disruption. We used the parameters presented in Sect. 4.1, and set the mutation rate $\mu = 0.6$.

Table 1 shows the results of NSGA-II within a time limit of 300 s. For each evaluation model (shown in row "Eval.") and operator (shown in row "Operator") combination, we give the average hypervolume of the non-dominated fronts $\mathcal{F}_1$ we obtained (HV_{avg}), and the hypervolume on the best run (HV_{max}). In the last column "$\epsilon - 300$", we indicate the hypervolume of the front obtained using the ϵ-constraint method with the model introduced in [13], with a 300-second time limit. Due to the time limit, this method does not always yield the full optimal Pareto front.

Table 1. Average hypervolume of fronts obtained by NSGA-II in 300 s.

Eval	Asset penalization				Assignment penalization				$\epsilon - 300$
Operator	Single-change		Multi-change		Single-change		Multi-change		
	HV_{max}	HV_{avg}	HV_{max}	HV_{avg}	HV_{max}	HV_{avg}	HV_{max}	HV_{avg}	HV
n = 30	84.3%	79.9%	84.2%	80.8%	85.0%	80.5%	84.7%	81.2%	86.3%
40	82.2%	76.7%	82.9%	79.0%	81.6%	77.0%	83.4%	79.7%	82.5%
50	78.9%	74.6%	80.2%	76.7%	80.5%	76.1%	81.7%	77.9%	70.6%
60	78.4%	73.0%	78.7%	72.5%	80.1%	74.4%	79.9%	73.9%	58.9%

We can see that the second evaluation model on average gave fronts with higher hypervolume on average than the first model for the same operators. For instances with 30, 40 and 50 assets, the multi-change operator performed better than the single-change operators. The multi-change operator offers more stable results than the single-change operators, and find solutions with higher profit. For larger instances, we obtain better fronts on average than the $\epsilon - constraint$ method within the same time limit.

Based on Table 1, we will consider the second evaluation model with multi-change operator to evaluate our additional components. We performed a parameter analysis similar to Sect. 4.1 to determine good values for our adaptive method and local search parameters. We set the initial population to $N = 50$, and offspring population to $M = 50/4$. For the local search, we apply it to 5% of the solutions, with a relative gap tolerance of 0.05. Each model is run five times on each instance, to ensure the robustness of our results. In order to avoid overfitting, we selected new breakdowns at random for the benchmark instances.

Table 2 summarizes the results of our algorithm with no additional component, with the adaptive parameters enabled and our local search procedure. It shows the average value of the hypervolume found in the best run (HV_{max}) and the average value in all the runs (HV_{avg}).

Table 2. Comparison of the components of our NSGA-II implementation.

Method	No component		Adaptive		Local Search		$\epsilon - 300$
	HV_{max}	HV_{avg}	HV_{max}	HV_{avg}	HV_{max}	HV_{avg}	HV
n = 30	79.5%	77.6%	79.8%	77.8%	82.3%	81.9%	82.3%
40	75.8%	73.5%	75.8%	73.6%	79.9%	79.2%	79.1%
50	74.5%	72.2%	75.1%	73.1%	80.7%	79.5%	68.9%
60	70.5%	67.8%	73.0%	69.7%	80.0%	78.3%	56.4%

The adaptive component yielded similar results for instances with 30 and 40 assets and slightly better results for 50 and 60 assets when enabled. We obtained significant improvements for all instances when enabling our local search procedure, up 10% for instances with 60 assets on average. The local search procedure also improved the stability of our algorithm, reducing the gap between the best solution and the average solution for all size of instances.

5   Conclusion

NSGA-II is a popular algorithm for multi-objective heuristic resolution that has proven efficient for multiple vehicle routing problems. We proposed an implementation of NSGA-II for the D-APP. Due to the numerous constraints that are part of the D-APP, we introduced mutation and crossover operators based on properties of the problem and MIPs to repair and evaluate solutions. It is the first heuristic solution method dedicated to the D-APP. The approach can be improved by defining operators that better take the deviation into account, and finding faster procedures to repair and evaluate a solution.

Acknowledgment. This study was carried out within the framework of GEOSAFE (Geospatial Based Environment For Optimization Systems Addressing Fire Emergencies). This work was partially supported by the framework of the Labex MS2T, funded by the French Government, via the program "Investments for the future" managed by the National Agency for Research (Reference ANR-11-IDEX-0004-02).

References

1. Ben-Said, A., Moukrim, A., Guibadj, R.N., Verny, J.: Using decomposition-based multi-objective algorithm to solve selective pickup and delivery problems with time windows. Comput. Oper. Res. **145**, 105867 (2022). https://doi.org/10.1016/j.cor.2022.105867

2. Cheng, J., Zhang, G., Li, Z., Li, Y.: Multi-objective ant colony optimization based on decomposition for bi-objective traveling salesman problems. Soft. Comput. **16**, 597–614 (2012). https://doi.org/10.1007/s00500-011-0759-3

3. Deb, K., Pratap, A., Agarwal, S., Meyarivan, T.: A fast and elitist multiobjective genetic algorithm: NSGA-II. IEEE Trans. Evol. Comput. **6**(2), 182–197 (2002). https://doi.org/10.1109/4235.996017

4. Jemai, J., Zekri, M., Mellouli, K.: An NSGA-II algorithm for the green vehicle routing problem. In: Hao, J.-K., Middendorf, M. (eds.) EvoCOP 2012. LNCS, vol. 7245, pp. 37–48. Springer, Heidelberg (2012). https://doi.org/10.1007/978-3-642-29124-1_4

5. Jozefowiez, N., Semet, F., Talbi, E.-G.: Enhancements of NSGA II and its application to the vehicle routing problem with route balancing. In: Talbi, E.-G., Liardet, P., Collet, P., Lutton, E., Schoenauer, M. (eds.) EA 2005. LNCS, vol. 3871, pp. 131–142. Springer, Heidelberg (2006). https://doi.org/10.1007/11740698_12

6. Jozefowiez, N., Semet, F., Talbi, E.G.: Multi-objective vehicle routing problems. Eur. J. Oper. Res. **189**(2), 293–309 (2008). https://doi.org/10.1016/j.ejor.2007.05.055

7. Mas-Colell, A., Whinston, M.D., Green, J.R., et al.: Microeconomic Theory, vol. 1. Oxford University Press, New York (1995)

8. Merwe, M.: An optimisation approach for assigning resources to defensive tasks during wildfires. RMIT University (2015). https://researchbank.rmit.edu.au/view/rmit:161622

9. Merwe, M., Minas, J., Ozlen, M., Hearne, J.: A mixed integer programming approach for asset protection during escaped wildfires. Can. J. For. Res. **45**, 04 (2015). https://doi.org/10.1139/cjfr-2014-0239

10. Merwe, M., Ozlen, M., Hearne, J., Minas, J.: Dynamic rerouting of vehicles during cooperative wildfire response operations. Ann. Oper. Res. **254**, 07 (2017). https://doi.org/10.1007/s10479-017-2473-8

11. Mirzaei, M.H., Ziarati, K., Naghibi, M.-T.: Bi-objective version of team orienteering problem (BTOP). In: 2017 7th International Conference on Computer and Knowledge Engineering (ICCKE), pp. 1–7 (2017). https://doi.org/10.1109/ICCKE.2017.8167930

12. Pareto, V.: Manual of Political Economy. Augustus M. Kelley Publishers, New York (1971)

13. Peña, Q., Serairi, M., Moukrim, A.: Reformulation and valid inequalities for the dynamic asset protection problem during an escaped wildfire (2022)

14. Schilde, M., Doerner, K., Hartl, R., Kiechle, G.: Metaheuristics for the bi-objective orienteering problem. Swarm Intell. **3**, 179–201 (2009). https://doi.org/10.1007/s11721-009-0029-5

15. Zitzler, E., Thiele, L.: Multiobjective evolutionary algorithms: a comparative case study and the strength Pareto approach. IEEE Trans. Evol. Comput. **3**(4), 257–271 (1999). https://doi.org/10.1109/4235.797969

Author Index

Printed in the United States
by Baker & Taylor Publisher Services

Printed in the United States
by Baker & Taylor Publisher Services